6th Social Media Mining for Health Workshop and Shared Tasks (SMM4H 2021)

Online
10 June 2021

ISBN: 978-1-7138-3029-0

Printed from e-media with permission by:

Curran Associates, Inc.
57 Morehouse Lane
Red Hook, NY 12571

CURRAN ASSOCIATES INC. proceedings .com

Some format issues inherent in the e-media version may also appear in this print version.

Association for Computational Linguistics
209 N. Eighth Street
Stroudsburg, Pennsylvania 18360

Phone: 1-570-476-8006
Fax: 1-570-476-0860

acl@aclweb.org

Additional copies of this publication are available from:

Curran Associates, Inc.
57 Morehouse Lane
Red Hook, NY 12571 USA
Phone: 845-758-0400
Fax: 845-758-2633
Email: curran@proceedings.com
Web: www.proceedings.com

NAACL-HLT 2021

Social Media Mining for Health (SMM4H)

Proceedings of the Sixth Workshop and Shared Tasks

June 10, 2021

Preface

Welcome to the 6th Social Media Mining for Health (#SMM4H) Workshop & Shared Task 2021, co-located at the 2021 Annual Conference of the North American Chapter of the Association for Computational Linguistics. Held online in its sixth iteration, #SMM4H 2021 continues to serve as a venue for bringing together data mining researchers interested in building solutions for challenges involved in utilizing social media data for health informatics. For #SMM4H 2021, we accepted 3 workshop papers and 29 shared task system description papers. Each submission was peer-reviewed by two to three reviewers.

The accepted workshop papers used social media data, mainly from Twitter, for topics, studies and applications surrounding COVID-19 and pharmacovigilance. Niu et. al. present a study summarizing the evaluation of Twitter sentiments towards non-pharmaceutical interventions for COVID-19 in Canada. Karisani et. al. propose a novel technique that uses both unlabeled and labeled tweets with drug mentions along multiple views to achieve a new state-of-the-art performance in extracting adverse drug effects. Finally, Miranda et. al. present a new annotated corpora in Spanish to identify occupational subgroups on Twitter to estimate risks associated with COVID-19. They also present a summary of the ProfNER shared task organized with the annotated data along the text classification and named entity recognition subtasks.

The #SMM4H 2021 shared tasks sought to advance the use of Twitter data (tweets) for pharmacovigilance, medication non-adherence, patient-centered outcomes, tracking cases and symptoms associated with COVID-19 and assessing risks for occupational groups. In addition to re-reruns of adverse drug effects extraction tasks in English and Russian #SMM4H 2021 included new tasks for detecting medication non-adherence, adverse pregnancy outcomes, probable cases of COVID-19, symptoms associated with COVID-19, extracting occupations and professions from Spanish tweets for COVID-19 risk assessment and detecting self reports of breast cancer posts. The eight tasks required methods for binary classification, multi-class classification, and named entity recognition (NER). With 40 teams making prediction submissions, participation in the #SMM4H shared tasks continue to grow. Among the 29 shared task system description papers that were accepted, 9 teams were invited to present their system orally.

The organizing committee of #SMM4H 2021 would like to thank the program committee for reviewing the workshop papers and the additional reviewers of system description papers for providing constructive feedback and participating in peer-review. We are also grateful to the organizers of NAACL 2021 for facilitating the organization of the workshop and the Codalab team for providing the platform to organize shared tasks. We would also like to thank the annotators of the shared task datasets, and of course, everyone who submitted a paper or participated in the shared tasks. #SMM4H 2021 would not have been possible without the contributions and participation from all of them.

Arjun, Ari, Antonio, Mohammed Ali, Ilseyar, Zulfat, Eulàlia, Salvador, Ivan, Karen, Davy, Elena, Abeed, Juan, Martin and Graciela

Organizing and Program Committees

Organizing Committee:

Graciela Gonzalez-Hernandez, University of Pennsylvania, USA
Arjun Magge, University of Pennsylvania, USA
Ari Z. Klein, University of Pennsylvania, USA
Davy Weissenbacher, University of Pennsylvania, USA
Ivan Flores, University of Pennsylvania, USA
Karen O'Connor, University of Pennsylvania, USA
Martin Krallinger, Barcelona Supercomputing Center, Spain
Antonio Miranda-Escalada, Barcelona Supercomputing Center, Spain
Eulàlia Farré-Maduell, Barcelona Supercomputing Center, Spain
Salvador Lima López, Barcelona Supercomputing Center, Spain
Juan M. Banda, Georgia State University, USA
Abeed Sarker, Emory University, USA
Mohammed Ali Al-garadi, Emory University, USA
Elena Tutubalina, Kazan Federal University, Russia
Ilseyar Alimova, Kazan Federal University, Russia
Zulfat Miftahutdinov, Kazan Federal University, Russia

Program Committee:

Olivier Bodenreider, US National Library of Medicine, USA
Pierre Zweigenbaum, French National Center for Scientific Research, France
Kirk Roberts, University of Texas Health Science Center at Houston, USA
Rajesh Piryani, South Asian University, India
Yutaka Sasaki, Toyota Technological Institute, Japan
Nicolas Turenne, French National Institute for Agricultural Research, France

Table of Contents

Conference Program

June 10th 2021

09:00–09:15 *Opening Remarks and Introduction*
Graciela Gonzalez-Hernandez

09:15–10:15 Oral Presentations Q&A Session 1

Statistically Evaluating Social Media Sentiment Trends towards COVID-19 Non-Pharmaceutical Interventions with Event Studies
Jingcheng Niu, Erin Rees, Victoria Ng and Gerald Penn

View Distillation with Unlabeled Data for Extracting Adverse Drug Effects from User-Generated Data
Payam Karisani, Jinho D. Choi and Li Xiong

Overview of the Sixth Social Media Mining for Health Applications (#SMM4H) Shared Tasks at NAACL 2021
Arjun Magge, Ari Klein, Antonio Miranda-Escalada, Mohammed Ali Al-Garadi, Ilseyar Alimova, Zulfat Miftahutdinov, Eulàlia Farré-Maduell, Salvador Lima López, Ivan Flores, Karen O'Connor, Davy Weissenbacher, Elena Tutubalina, Abeed Sarker, Juan Banda, Martin Krallinger and Graciela Gonzalez-Hernandez

The ProfNER shared task on automatic recognition of occupation mentions in social media: systems, evaluation, guidelines, embeddings and corpora
Antonio Miranda-Escalada, Eulàlia Farré-Maduell, Salvador Lima López, Luis Gascó, Vicent Briva-Iglesias, Marvin Agüero-Torales and Martin Krallinger

10:15–10:30 Break

10:30–11:10 ***Invited Talk***
Mark Dredze

11:10–11:25 Break

June 10th 2021 (continued)

11:30–12:30 Oral Presentations Q&A Session 2

BERT based Transformers lead the way in Extraction of Health Information from Social Media
Sidharth Ramesh, Abhiraj Tiwari, Parthivi Choubey, Saisha Kashyap, Sahil Khose, Kumud Lakara, Nishesh Singh and Ujjwal Verma

KFU NLP Team at SMM4H 2021 Tasks: Cross-lingual and Cross-modal BERT-based Models for Adverse Drug Effects
Andrey Sakhovskiy, Zulfat Miftahutdinov and Elena Tutubalina

Transformer-based Multi-Task Learning for Adverse Effect Mention Analysis in Tweets
George-Andrei Dima, Dumitru-Clementin Cercel and Mihai Dascalu

Pre-trained Transformer-based Classification and Span Detection Models for Social Media Health Applications
Yuting Guo, Yao Ge, Mohammed Ali Al-Garadi and Abeed Sarker

12:30–13:15 Poster Session

13:15–13:30 Break

13:30–14:45 Oral Presentations Q&A Session 3

BERT Goes Brrr: A Venture Towards the Lesser Error in Classifying Medical Self-Reporters on Twitter
Alham Fikri Aji, Made Nindyatama Nityasya, Haryo Akbarianto Wibowo, Radityo Eko Prasojo and Tirana Fatyanosa

UACH-INAOE at SMM4H: a BERT based approach for classification of COVID-19 Twitter posts
Alberto Valdes, Jesus Lopez and Manuel Montes

System description for ProfNER - SMMH: Optimized finetuning of a pretrained transformer and word vectors
David Carreto Fidalgo, Daniel Vila-Suero, Francisco Aranda Montes and Ignacio Talavera Cepeda

Word Embeddings, Cosine Similarity and Deep Learning for Identification of Professions & Occupations in Health-related Social Media
Sergio Santamaría Carrasco and Roberto Cuervo Rosillo

June 10th 2021 (continued)

	Classification, Extraction, and Normalization : CASIA_Unisound Team at the Social Media Mining for Health 2021 Shared Tasks Tong Zhou, Zhucong Li, Zhen Gan, Baoli Zhang, Yubo Chen, Kun Niu, Jing Wan, Kang Liu, Jun Zhao, Yafei Shi, Weifeng Chong and Shengping Liu
14:45–15:00	*Conclusion and Closing Remarks* Graciela Gonzalez-Hernandez

Statistically Evaluating Social Media Sentiment Trends towards COVID-19 Non-Pharmaceutical Interventions with Event Studies

Jingcheng Niu[1,2,3] **Erin E. Rees**[1] **Victoria Ng**[1] **Gerald Penn**[2,3]
[1]Public Health Agency of Canada [2]University of Toronto [3]Vector Institute
{niu,gpenn}@cs.toronto.edu
{erin.rees,victoria.ng}@canada.ca

Abstract

In the midst of a global pandemic, understanding the public's opinion of their government's policy-level, non-pharmaceutical interventions (NPIs) is a crucial component of the health-policy-making process. Prior work on COVID-19 NPI sentiment analysis by the epidemiological community has proceeded without a method for properly attributing sentiment changes to events, an ability to distinguish the influence of various events across time, a coherent model for predicting the public's opinion of future events of the same sort, nor even a means of conducting significance tests. We argue here that this urgently needed evaluation method does already exist. In the financial sector, *event studies* of the fluctuations in a publicly traded company's stock price are commonplace for determining the effects of earnings announcements, product placements, etc. The same method is suitable for analysing temporal sentiment variation in the light of policy-level NPIs. We provide a case study of Twitter sentiment towards policy-level NPIs in Canada. Our results confirm a generally positive connection between the announcements of NPIs and Twitter sentiment, and we document a promising correlation between the results of this study and a public-health survey of popular compliance with NPIs.

1 Introduction

As COVID-19 spreads rapidly around the world, governments have implemented different NPIs to contain the spread of the virus. While effective at slowing down the spread of COVID-19 (Haug et al., 2020), NPIs such as school and non-essential businesses closures, telecommuting, mask requirements and physical distancing measures have drastically changed our lives and sparked dissent. Anti-mask and anti-lockdown protests are commonplace, while there are nearly fifty million active cases around the world. It is crucial for decision makers to understand the public's opinion about NPIs, and for policy-makers to have a means of forecasting the level of popular compliance with them. This will determine their effectiveness as well as whether additional measures and communication strategies are needed in light of waning adherence.

Analysis of social media data is already popular among epidemiologists, as it is a data source with near real-time feedback at very low cost (Majumder et al., 2016). Extracting sentiment trends towards the pandemic on various social media platforms has already attracted interest (Wang et al., 2020b; Li et al., 2020; Wang et al., 2020a). Neural sentiment analysis is very prevalent because of its high performance on classification tasks[1] and versatility. Temporal variation of sentiment is usually represented by time series, in which an average model-predicted sentiment scores over from all social media posts within each time interval is computed. Previous work following this paradigm suffers from two major issues, however.

Firstly, nearly all time-series analyses have been based on sentiment classification results — every post is classified into one of the predetermined sentiment categories (positive/(neutral)/negative) — even though sentiment is a continuous random variable. For example, Wang et al. (2020b) provide two "sentiment-neutral" examples that in fact have differing sentiments. Smoothing sentiment from a continuous variable into a ternary or binary scale causes a loss of dynamics, hence increasing the difficulty of the task and lowering the reliability of all subsequent analyses. There are now n-valued sentiment corpora for $n = 5$ (Socher et al., 2013) and $n = 7$ (Mohammad et al., 2018), but finer-grained discrete sentiment does not entirely solve the problem. The valence regression task (*V-reg*) proposed by Mohammad et al. (2018) is far more suitable because it conveys a continuous sentiment intensity measure through a logistic regression score.

[1]Top performers achieve near perfect accuracies, e.g., Jiang et al. (2020) at 97.5%.

Proceedings of the Sixth Social Media Mining for Health Workshop 2021, pages 1–6
June 10, 2021.

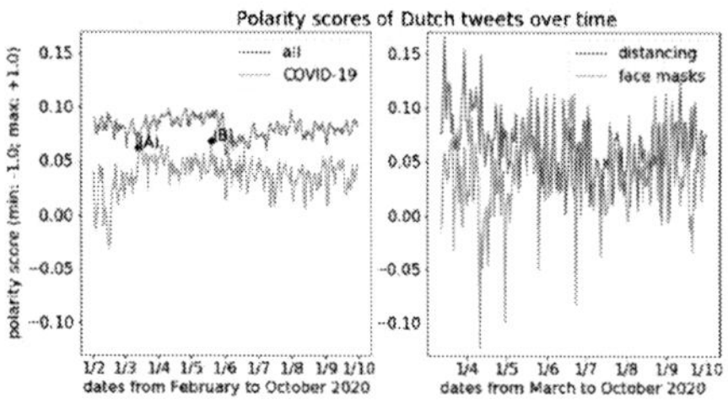

Figure 1: Wang et al. (2020a) claimed that general sentiment reached a minimum when the government announced a "lock-down" (A), and COVID-19 related sentiment reached a maximum when Amsterdam announced release measures (B). Note that the magnitude of difference between the minimum point they discovered at (A) and the valley a few days prior, at which there was no press conference, is not visible to the naked eye.

A continuous score also allows us to compute an average sample sentiment over a definite period of time, which has a more accurate variance than smoothing binary scores.

Secondly, because of the community's lack of a model capable of conducting significance tests and distinguishing the influence of various events across time, no statistically sound conclusion can be drawn. As an example, Wang et al. (2020a) claimed to have noticed a link between public sentiment and the timing of the Dutch government's press conferences by visually inspecting the raw trend of social media sentiment, seen in Figure 1. In fact, there were numerous peaks and valleys throughout the interval they studied, because the average sentiment fluctuated wildly during this time.

We can bring the potential of this urgently needed application to fruition by looking outside CL/NLP. Financial analysts face similar problems when they try to assess the effect of a particular news event on the price of a particular stock, because the price is affected by countless events as well as the reactions of traders with different motivations and perspectives on those events. *Event studies* (Brown and Warner, 1980, 1985) have been proposed and recognised as viable methods for attributing stock price fluctuations to specific financial events. To our knowledge, there has been no study of this class of methods within epidemiology.

2 Event Attribution

2.1 In Finance

In the financial sector, event studies are used to examine the return behaviour of a security after the market experiences some event (e.g., a stock split or an earnings release) that pertains to the firm that issued the security. The actual return of a stock (or a portfolio of assets) (R_t) at a given time t ($t = 0$ represents the time of the event) can be decomposed as follows: $R_t = \mathbb{E}[R_t|X_t] + \xi_t$. $\mathbb{E}[R_t|X_t]$ is an expected return, which can be explained by a model given the conditioning information X_t. ξ_t is an "abnormal" return that directly measures the unexpected changes on the returns, which are likely to have been caused by some unforeseen event (Eckbo, 2009). It is also possible that the abnormal return was just caused by chance ($\mathbb{E}[\xi_t] = 0$), however, and we can measure the statistical significance with which we can reject this null hypothesis through various tests based upon *time-series aggregation*, which we discuss presently.

The expected return can be estimated by a *market model* (Fama and MacBeth, 1973): $\mathbb{E}[R_t|X_t] = \alpha + \beta R_{m,t}$, where $R_{m,t}$ is the return of a market portfolio, i.e., of all of the assets in the market as represented by a broad market index (e.g., S&P 500, Nasdaq). β is the risk factor of the stock and can be computed using the ratio of the covariance between the actual return and the market return to the variance of the market return $\beta = \frac{cov(R, R_m)}{\sigma^2(R_m)}$. α is the bias that can be computed with least squares estimation, but since β is already computed, the optimal value of α is $\frac{1}{N}\sum_t^N(R_t - \beta R_{m,t})$ where N is the sample size.

The analysis of an event proceeds by first determining whether there is a statistically significant impact, and then if there is, computing the magnitude of the impact. To answer these two questions, the integral of the abnormal return, called the *cumulative average residual (CAR)*, is computed: $\text{CAR}(t_1, t_2) = \sum_{t=t_1}^{t_2} \xi_t$. Under the assumption that the return of a stock with no marked events is a stochastic process that perfectly reflects the overall performance of the market as accounted for by the market model (Fama and MacBeth, 1973), the expectation of CAR should be zero. Thus, we can test the null hypothesis that the event has no impact on the return, $\mathbb{E}[\xi_t] = 0$, by a one-sample t-test, one-sample Wilcoxon signed rank test (Wilcoxon, 1945), or a binomial proportionality z-test. In finance, the ratio of CAR divided by the overall actual return is traditionally used to represent the magnitude of an event's impact, but the statistics of these tests can also be used.

2.2 In Public Health

Over the course of the pandemic, governments around the world have utilized different NPIs at different times and with different stringencies (Hale et al., 2020). Therefore, overall sentiment shift cannot represent the impact of individual public health events. Instead, overall sentiment acts like market return: an aggregation of individual sentiments. Therefore, we define the daily sentiment index (I) as the average sentiment (valence) of all the tweets from a single day. Individual COVID-19-related topics are analogous to individual stocks, and the sentiment change on individual topics is reflected in the change of the sentiment index. But some topics specifically relate to certain events, similar to how individual stocks react to the news relevant to their firms. Therefore, the average sentiment $S_{m,t}$ of all discussions on topic m at time t is similar to the return of a stock in the event study. Our "market model" for sentiment is: $\mathbb{E}[S_{m,t}] = \alpha_m + \beta_m I_t$. We compute the abnormal sentiment by $\xi_{m,t} = S_{m,t} - \mathbb{E}[S_{m,t}]$ and calculate CAR by aggregating $\xi_{m,t}$ over time: $\text{CAR}(t_1, t_2) = \sum_{t=t_1}^{t_2} \xi_t$.

3 Experimental Setup

Gilbert et al. (2020) started collecting COVID-19 related tweets by searching for tweets mentioning at least one of the various naming conventions for COVID-19 using the Twitter search API as at January 21, 2020, and collected 281,487,148 tweets up until August 23rd, 2021. After Carmen geolocation (Dredze et al., 2013), we obtained 5,979,759 English Twitter samples from Canada.

For this paper, we studied two NPIs: wearing a mask and social distancing. For present purposes, we considered an event to be every change in the stringency level of any NPI, as measured by the Oxford COVID-19 Government Response Tracker (OxCGRT) project (Hale et al., 2020). We used a keyword-based filter to obtain topic-related tweets. We began with a manually written list of related keywords to obtain a list of tweets M that contain a keyword, and $\overline{M}$ that do not contain any keyword. Then for each bigram and trigram x, we calculated a topic relevance score based on pointwise mutual information: $pmi(x; M) - pmi(x; \overline{M})$. We ranked the top 150 keywords for each n-gram and manually removed the topic-unrelated ones. For example, "covidsafe" was identified using this method but "congressman sponsor," a topic relevance score of 14.59, was nevertheless manually removed.

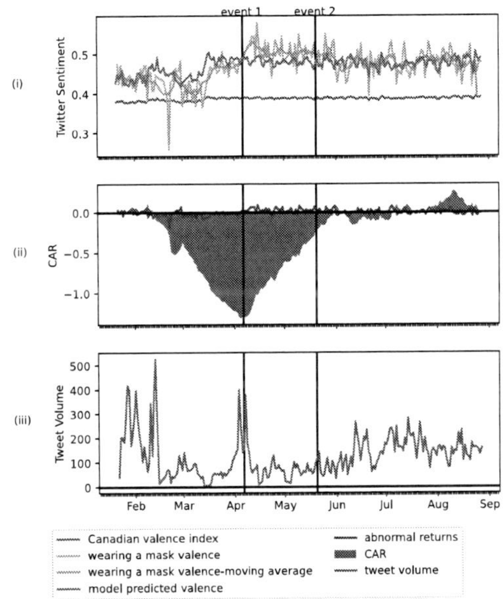

(a) Wearing a mask sentiment analysis

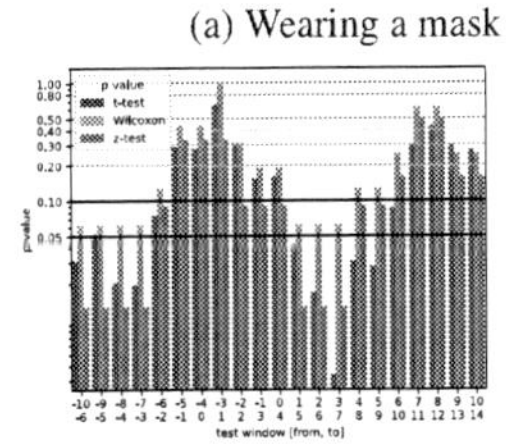

(b) Event 1 significance study

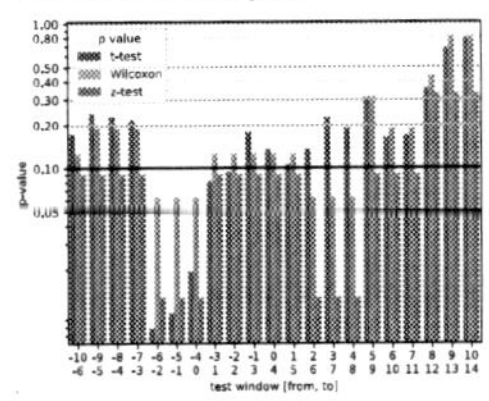

(c) Event 2 significance study

Figure 2: Wearing a mask event significance

After filtering all the tweets connected to an NPI of interest, we computed their valence score using the NTUA-SLP model,[2] which was selected from the 75 entrants to the V-reg shared task (Mohammad et al., 2018). We followed the hyperparameter settings from the original paper (Baziotis et al., 2018) and reproduced its reported Pearson correlation (0.846) on the English valence dataset. To establish a periodic time series of valence change, we computed the daily average valence of tweets posted on the same day.[3]

4 Individual NPIs Experimental Results

Wearing A Mask Canada's mask advisory has changed several times during the progression of the pandemic (Mohammed et al., 2020) and we investigated two key changing points of the advisory

[2]https://github.com/cbaziotis/ntua-slp-semeval2018

[3]Our subsequent analyses and data are publicly available: https://github.com/frankniujc/covid_sentiment_analysis.

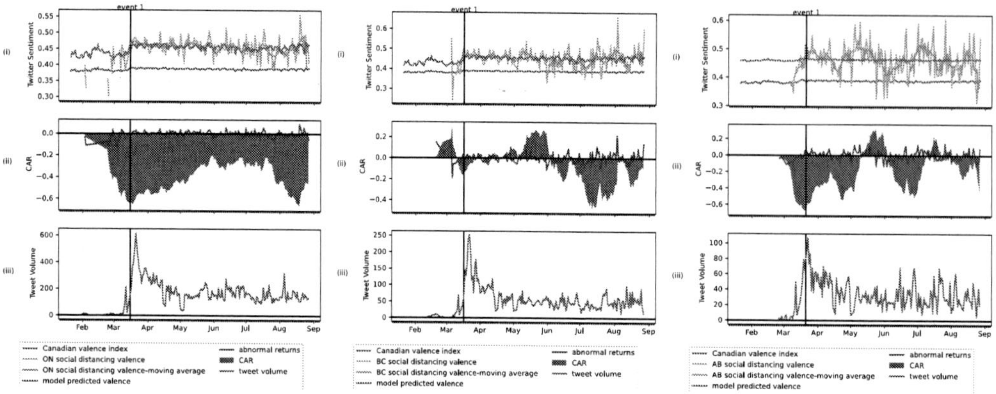

(a) Ontario (initial: Mar 16) (b) British Columbia (initial: Mar 17) (c) Alberta (initial: Mar 21)

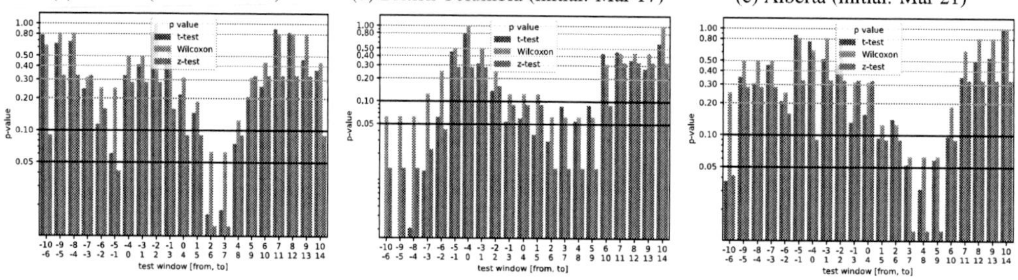

(d) ON significance (e) BC significance (f) AB significance

Figure 3: Social distancing recommendation event significance by province.

as events. On April 6th, 2020, the Public Health Agency of Canada (PHAC) revised the advisory for mask wearing (event 1), permitting the use of non-medical face coverings in public (Chase, 2020; Mohammed et al., 2020). Finally on May 20, 2020, PHAC formally issued a recommendation for the general public to wear masks in public (event 2) (Mohammed et al., 2020; Harris, 2020).

Assuming a confidence threshold of $\alpha = 0.05$, event 1 had a statistically significant positive impact for up to 9 days (Figure 2b). Event 2 also showed significance from two days after the event to up to eight days after ($[+2, +8]$; Figure 2c). Unlike event 1, there is also a period of significance right before the event occurred. This may have been anticipatory, or it may indicate that the observed impact had instead been caused by prior events. During the 9-day effect window of event 1, there is a 2.13% positive CAR, with t-statistic 1.73, Wilcoxon statistic 7.0, and z-statistic 1.67.

Social Distancing Social distancing recommendations have been issued with different stringencies and at different times at the provincial level in Canada. Therefore, we focus separately on three provinces: Ontario (ON), British Columbia (BC) and Alberta (AB), with sufficient numbers of tweets and different distancing policies. According to McCoy et al. (2020), Ontario released its first province-wide social distancing recommendation on March 16, 2020 (Williams, 2020); British Columbia issued a social distancing recommendation on March 17, 2020 (Dix and Henry, 2020); and lastly, Alberta released a public message about social distancing on March 21st[4] (McCoy et al., 2020).

Figure 3 analyses the significance of the initial recommendations in those three provinces. All three announcements have a positive impact on CAR with statistical significance. Ontario's recommendation (Figure 3d) has a short but significant impact on $[+2, +7]$. Alberta (Figure 3f) exhibits a significant impact on $[+3, +9]$, and British Columbia on $[+1, +9]$.

5 CAR and Survey Data Correlation

To help understand whether the sentiment of NPIs measured using Twitter are representative of the general Canadian population, we assessed the correlation between our NPI sentiments and the level of compliance measured through a national survey.

[4]https://www.alberta.ca/prevent-the-spread.aspx

The COVID-19 Monitor initiative (COV, 2020; Mohammed et al., 2020) has conducted 25 surveys in Canada on people's compliance with 6 NPIs since mid-March. Each survey has approximately 2000 participants. The demographics of the participants have been pre-stratified, and each wave was post-stratified by modelling raking weights based on the 2010 Canadian Census. Among the 6 NPIs, both *social distancing* and *wearing a mask* appear. For the cross-correlation test, both time series have been detrended using the SciPy signal package,[5] and then pre-whitened following the instructions proposed by Dean and Dunsmuir (2016) to remove autocorrelations with the time series.[6] Figure 4 shows the correlations and cross-correlations with the proportion of the population who report complying with either of these two NPIs and CAR. Wearing a mask receives a strong Pearson $r = 0.915$ (Figure 4a), a cross-correlation of 0.710 and a +5 lag, meaning CAR is 5 days ahead of the survey (Figure 4b). Social distancing receives a moderate Pearson $r = 0.481$ (Figure 4c), a cross-correlation of 0.492 and also a +5 lag (Figure 4d). The cross-correlations cannot be quantitatively compared with the Pearson correlation scores as they are calculated differently, but the general trend stays the same: wearing a mask exhibits a strong correlation while social distancing, only moderate one. The lags also accord with our expectations as COV (2020) conducted surveys 4 to 10 days apart.

The lower correlation for social distancing might have been caused by their more diverse implementation across subsovereign jurisdictions (see section 4). As the details of the sample selection process at the provincial level are not publicly available, we have not been able to draw direct, provincial comparisons. Mask-wearing advisories, however, are mostly issued at the federal level in Canada. Comparing mask-wearing across provinces is thus less problematic. With both types of NPI, Twitter users are demographically younger, better educated, and more urban than the general population (Mellon and Prosser, 2017; Murthy et al., 2016). This may explain some differences from the national distribution sampled for this survey.

[5]https://docs.scipy.org/doc/scipy/reference/signal.html

[6]We tested the autocorrelation of both the CARs and survey data. The level of autocorrelation in all the time series is low, and applying pre-whitening did not result in different conclusions in this study.

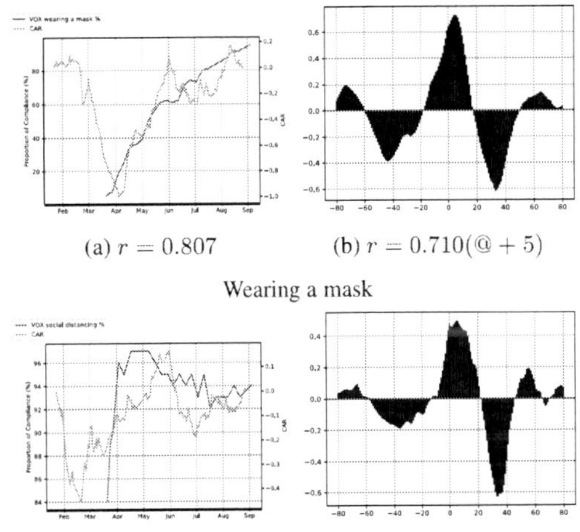

(a) $r = 0.807$ (b) $r = 0.710(@ + 5)$

Wearing a mask

(c) $r = 0.439$ (d) $r = 0.492(@ + 5)$

Social distancing

Figure 4: CAR and compliance survey correlation. Captions of (a) and (c) report Pearson correlations; captions of (b) and (d) report cross-correlations with days of lag.

Acknowledgements

This study is funded by the Canadian Safety and Security Program (CSSP) from Defence Research and Development Canada (DRDC). Many thanks for technical support and study design guidance to Jean-Philippe Gilbert (Laval University), Hélène Carabin, Esther Perez, Mireille D'Astous, Simon de Montigny, and Nasri Bouchra (University of Montreal), Patrick Daley (Public Health Agency of Canada), and Suzanne Hindmarch (University of New Brunswick). We also thank Saif Mohammad (National Research Council of Canada) for his help with sentiment analysis, and Tong Lin (University of Toronto) for his help with event studies.

References

2020. COVID-19 Monitor. Technical report, Vox Pop Labs.

C. Baziotis, A. Nikolaos, A. Chronopoulou, A. Kolovou, G. Paraskevopoulos, N. Ellinas, S. Narayanan, and A. Potamianos. 2018. NTUA-SLP at SemEval-2018 Task 1: Predicting Affective Content in Tweets with Deep Attentive RNNs and Transfer Learning. In *Proceedings of The 12th International Workshop on Semantic Evaluation*, pages 245–255.

S. Brown and J. Warner. 1985. Using daily stock returns. *Journal of Financial Economics*, 14(1):3–31.

S. J. Brown and J. B. Warner. 1980. Measuring security price performance. *Journal of Financial Economics*, 8(3):205–258.

S. Chase. 2020. Theresa Tam offers new advice: Wear a non-medical face mask when shopping or using public transit. *CBC News*.

R. T. Dean and W. T. M. Dunsmuir. 2016. Dangers and uses of cross-correlation in analyzing time series in perception, performance, movement, and neuroscience: The importance of constructing transfer function autoregressive models. *Behavior Research Methods*, 48(2):783–802.

A. Dix and B. Henry. 2020. Joint statement on Province of B.C.'s COVID-19 response, latest updates | BC Gov News. *BC Gov News*.

M. Dredze, M. J. Paul, S. Bergsma, and H. Tran. 2013. Carmen: A twitter geolocation system with applications to public health. In *AAAI Workshop on Expanding the Boundaries of Health Informatics Using AI (HIAI)*.

B. E. Eckbo, ed. 2009. *Handbook of Corporate Finance: Empirical Corporate Finance. Vol. 1: ...*, reprinted edition. Number 3 in Handbooks in Finance. Elsevier North-Holland, Amsterdam.

E. F. Fama and J. D. MacBeth. 1973. Risk, Return, and Equilibrium: Empirical Tests. *Journal of Political Economy*, 81(3):607–636.

J.-P. Gilbert, J. Niu, S. de Montigny, V. Ng, and E. E. Rees. 2020. Machine Learning Identification of Self-Reported COVID-19 Symptoms from Tweets in Canada. In *AAAI 2021 W3PHIAI-21 Workshop*.

T. Hale, N. Angrist, E. Cameron-Blake, L. Hallas, B. Kira, S. Majumdar, A. Petherick, T. Phillips, H. Tatlow, and S. Webster. 2020. Variation in government responses to COVID-19.

K. Harris. 2020. Canadians should wear masks as an 'added layer of protection,' says Tam. *CBC News*.

N. Haug, L. Geyrhofer, A. Londei, E. Dervic, A. Desvars-Larrive, V. Loreto, B. Pinior, S. Thurner, and P. Klimek. 2020. Ranking the effectiveness of worldwide COVID-19 government interventions. *medRxiv* 2020.07.06.20147199.

H. Jiang, P. He, W. Chen, X. Liu, J. Gao, and T. Zhao. 2020. SMART: Robust and Efficient Fine-Tuning for Pre-trained Natural Language Models through Principled Regularized Optimization. In *Proceedings of the 58th Annual Meeting of the Association for Computational Linguistics*, pages 2177–2190.

S. Li, Y. Wang, J. Xue, N. Zhao, and T. Zhu. 2020. The Impact of COVID-19 Epidemic Declaration on Psychological Consequences: A Study on Active Weibo Users. *International Journal of Environmental Research and Public Health*, 17(6):2032.

M. S. Majumder, M. Santillana, S. R. Mekaru, D. P. McGinnis, K. Khan, and J. S. Brownstein. 2016. Utilizing Nontraditional Data Sources for Near Real-Time Estimation of Transmission Dynamics During the 2015-2016 Colombian Zika Virus Disease Outbreak. *JMIR Public Health and Surveillance*, 2(1):e30.

L. G. McCoy, J. Smith, K. Anchuri, I. Berry, J. Pineda, V. Harish, A. T. Lam, S. E. Yi, S. Hu, L. Rosella, B. Fine. 2020. Characterizing early Canadian federal, provincial, territorial and municipal nonpharmaceutical interventions in response to COVID-19: A descriptive analysis. *CMAJ Open*, 8(3):E545–E553.

J. Mellon and C. Prosser. 2017. Twitter and Facebook are not representative of the general population: Political attitudes and demographics of British social media users. *Research & Politics*, 4(3):2053168017720008.

S. Mohammad, F. Bravo-Marquez, M. Salameh, and S. Kiritchenko. 2018. SemEval-2018 Task 1: Affect in Tweets. In *Proceedings of The 12th International Workshop on Semantic Evaluation*, pages 1–17.

A. Mohammed, R. M. Johnston, and C. van der Linden. 2020. Public Responses to Policy Reversals: The Case of Mask Usage in Canada during COVID-19. *Canadian Public Policy*, 46(S2):S119–S126.

D. Murthy, A. Gross, and A. Pensavalle. 2016. Urban Social Media Demographics: An Exploration of Twitter Use in Major American Cities. *Journal of Computer-Mediated Communication*, 21(1):33–49.

R. Socher, A. Perelygin, J. Wu, J. Chuang, C. D. Manning, A. Ng, and C. Potts. 2013. Recursive Deep Models for Semantic Compositionality Over a Sentiment Treebank. In *Proceedings of the 2013 Conference on Empirical Methods in Natural Language Processing*, pages 1631–1642.

S. Wang, M. Schraagen, E. T. Kim Sang, and M. Dastani. 2020a. Public Sentiment on Governmental COVID-19 Measures in Dutch Social Media. In *Proceedings of the 1st Workshop on NLP for COVID-19 (Part 2) at EMNLP 2020*.

T. Wang, K. Lu, K. P. Chow, and Q. Zhu. 2020b. COVID-19 Sensing: Negative Sentiment Analysis on Social Media in China via BERT Model. *IEEE Access*, 8:138162–138169.

F. Wilcoxon. 1945. Individual Comparisons by Ranking Methods. *Biometrics Bulletin*, 1(6):80–83.

D. Williams. 2020. Enhanced Measures to Protect Ontarians from COVID-19. *Ontario Newsroom*.

View Distillation with Unlabeled Data for Extracting Adverse Drug Effects from User-Generated Data

Payam Karisani
Emory University
pkarisa@emory.edu

Jinho D. Choi
Emory University
jinho.choi@emory.edu

Li Xiong
Emory University
lxiong@emory.edu

Abstract

We present an algorithm based on multi-layer transformers for identifying Adverse Drug Reactions (ADR) in social media data. Our model relies on the properties of the problem and the characteristics of contextual word embeddings to extract two views from documents. Then a classifier is trained on each view to label a set of unlabeled documents to be used as an initializer for a new classifier in the other view. Finally, the initialized classifier in each view is further trained using the initial training examples. We evaluated our model in the largest publicly available ADR dataset. The experiments testify that our model significantly outperforms the transformer-based models pretrained on domain-specific data.

1 Introduction

Social media has made substantial amount of data available for various applications in the financial, educational, and health domains. Among these, the applications in healthcare have a particular importance. Although previous studies have demonstrated that the self-reported online social data is subject to various biases (Olteanu et al., 2018), this data has enabled many applications in the health domain, including tracking the spread of influenza (Aramaki et al., 2011), detecting the reports of the novel coronavirus (Karisani and Karisani, 2020), and identifying various illness reports (Karisani and Agichtein, 2018).

One of the well-studied areas in online public health monitoring is the extraction of adverse drug reactions (ADR) from social media data. ADRs are the unintended effects of drugs for prevention, diagnosis, or treatment. The researchers in Duh et al. (2016) reported that consumers, on average, report the negative effect of drugs on social media 11 months earlier than other platforms. This highlights the importance of this task. Another team of researchers in Golder et al. (2015) reviewed more than 50 studies and reported that the prevalence of ADRs across multiple platforms ranges between 0.2% and 8.0%, which justifies the difficulty of this task. In fact, despite the long history of this task in the research community (Yates and Goharian, 2013), for various reasons, the performance of the state-of-the-art models is still unsatisfactory. Social media documents are typically short and their language is informal (Karisani et al., 2015). Additionally, the imbalanced class distributions in ADR task has exacerbated the problem.

In this study we propose a novel model for extracting ADRs from Twitter data. Our model which we call View Distillation (VID) relies on the existence of two views in the tweets that mention drug names. We use unlabeled data to transfer the knowledge from the classifier in each view to the classifier in the other view. Additionally, we use a finetuning technique to mitigate the impact of noisy pseudo-labels after the initialization (Karisani and Karisani, 2021). As straightforward as it is to implement, our model achieves the state-of-the-art performance in the largest publicly available ADR dataset, i.e., SMM4H dataset. Our contributions are as follows: 1) We propose a novel algorithm to transfer knowledge across models in multi-view settings, 2) We propose a new technique to efficiently exploit unlabeled data in the supervised ADR task, 3) We evaluate our model in the largest publicly available ADR dataset, and show that it yields an additive improvement to the common practice of language model pretraining in this task. To our knowledge, our work is the first study that reports such an achievement. Next, we provide a brief overview of the related studies.

2 Related Work

Researchers have extensively explored the applications of ML and NLP models in extracting ADRs from user-generated data. Perhaps one of the early reports in this regard is published in Yates and

Proceedings of the Sixth Social Media Mining for Health Workshop 2021, pages 7–12
June 10, 2021.

Goharian (2013), where the authors utilize the related lexicons and extraction patterns to identify ADRs in user reviews. With the surge of neural networks in text processing, subsequently, the traditional models were aggregated with these techniques to achieve better generalization (Tutubalina and Nikolenko, 2017). The recent methods for extracting ADRs entirely rely on neural network models, particularly on multi-layer transformers (Vaswani et al., 2017).

In the shared task of SMM4H 2019 (Weissenbacher and Gonzalez-Hernandez, 2019), the top performing run was BERT model (Devlin et al., 2019) pretrained on drug related tweets. Remarkably, one year later in the shared task of SMM4H 2020 (Gonzalez-Hernandez et al., 2020), again a variant of pretrained BERT achieved the best performance (Liu et al., 2019). Here, we propose an algorithm to improve on pretrained BERT in this task. Our model relies on multi-view learning and exploits unlabeled data. To our knowledge, our model is the first approach that improves on the domain-specific pretrained BERT.

3 Proposed Method

Our model for extracting the reports of adverse drug effects rely on the properties of contextual neural word embeddings. Previous research on Word Sense Disambiguation (WSD) (Scarlini et al., 2020) has demonstrated that contextual word embeddings can effectively encode the context in which words are used. Although the representations of the words in a sentence are assumed to be distinct, they still possess shared characteristics. This is justified by the observation that the techniques such as self-attention (Vaswani et al., 2017), which a category of contextual word embeddings employ (Devlin et al., 2019), rely on the interconnected relations between word representations.

This property is particularly appealing when documents are short, therefore, word representations, if are adjusted accordingly, can be exploited to extract multiple representations for a single document. In fact, previous studies have demonstrated that word contexts can be used to process short documents, e.g., see the models proposed in Liao and Grishman (2011) and Karisani et al. (2020) for event extraction using hand-crafted features and contextual word embeddings respectively. Therefore, we use the word representations of drug mentions in user postings as the secondary view along the document representations of user postings in our model.

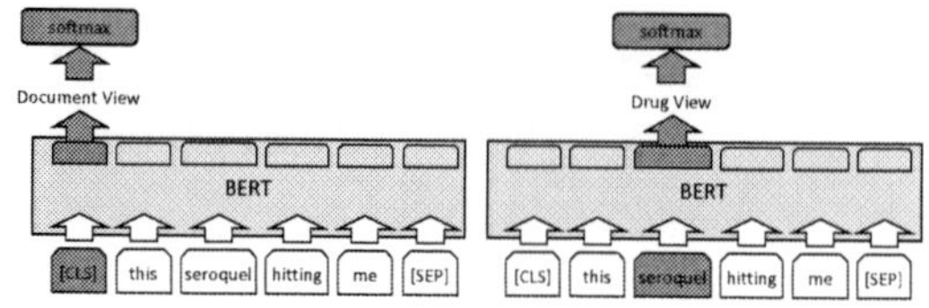

Figure 1: The illustration of the document and drug views in our model. We have used BERT as an encoder. See Devlin et al. (2019) for the format of input tokens.

As a concrete example, from the hypothetical tweet *"this seroquel hitting me"*, we extract one representation from the entire document and another representation from the drug name[1] Seroquel. In continue, we call these two views the document and drug views. Figure 1 illustrates these two views using BERT (Devlin et al., 2019) as an encoder.

Given the two views we can either concatenate the two sets of features and train a classifier on the resulting feature vector or use a co-training framework as described in Karisani et al. (2020). However, the former is not exploiting the abundant amount of unlabeled data, and the latter is resource intensive, because it is iterative, and also it has shown to be effective only in semi-supervised settings where there are only a few hundred training examples available. Therefore, below we propose an approach to effectively use the two views along the available unlabeled data in a supervised setting.

In the first step, we assume the classifier in each view is a student model and train this classifier using the pseudo-labels generated by the counterpart classifier. Since the labeled documents are already annotated, we carry out this step using the unlabeled documents. More concretely, let L and U be the sets of labeled and unlabeled user postings respectively. Moreover, let L_d and L_g be the sets of representations extracted from the document and drug views of the training examples in the set L; and let U_d and U_g be the document and drug representations of the training examples in the set U. To carry out this step, we train a classifier C_d on the representations in L_d and probabilistically, with temperature T in the softmax layer, label the representations in U_d. Then we use the association between the representations in U_d and U_g to construct a pseudo-labeled dataset of U_g. This dataset along its set of probabilistic pseudo-labels is used in a distillation technique (Hinton et al., 2015) to train a classifier called $\widehat{C_g}$. Correspondingly, we

[1]We assume every user posting contains only one drug name, in cases that there are multiple names we can use the first occurrence.

use the set L_g to train a classifier C_g, then label the set U_g and use the association between the data points in U_g and U_d to construct a pseudo-labeled dataset in the document view to train the classifier $\widehat{C_d}$.

The procedure above results in two classifiers $\widehat{C_d}$ and $\widehat{C_g}$. The classifier in each view is *initialized* by the knowledge transferred from the other view. However, the pseudo-labels that are used to train each classifier can be noisy. Thus, in order to reduce the negative impact of this noise, in the next step, we use the training examples in the sets L_d and L_g to further finetune these two classifiers respectively. To finetune $\widehat{C_d}$ we use the objective function below:

$$\mathcal{L}_d=\frac{1}{|L_d|}\sum_{v\in L_d}(1-\lambda)J(\widehat{C_d}(v),y_v)+\lambda J(\widehat{C_d}(v),C_d(v)), \quad (1)$$

where J is the cross-entropy loss, y_v is the ground-truth label of the training example v, and λ is a hyper-parameter to govern the impact of the two terms in the summation. The first term in the summation, is the regular cross-entropy between the output of $\widehat{C_d}$ and the ground-truth labels. The second term is the cross-entropy between the outputs of $\widehat{C_d}$ and C_d. We use the output of C_d as a regularizer to train $\widehat{C_d}$ in order to increase the entropy of this classifier for the prediction phase. Previous studies have shown that penalizing low entropy predictions increases generalization (Pereyra et al., 2017). We argue that this is particularly important in the ADR task, where the data is highly imbalanced. Note that, even though C_d is trained on the training examples in L_d, the output of this classifier for the training examples is not sparse–particularly for the examples with uncommon characteristics. Thus, we use these soft-labels[2] along the ground-truth labels to train $\widehat{C_d}$. Respectively, we use the objective function below to finetune $\widehat{C_g}$:

$$\mathcal{L}_g=\frac{1}{|L_g|}\sum_{v\in L_g}(1-\lambda)J(\widehat{C_g}(v),y_v)+\lambda J(\widehat{C_g}(v),C_g(v)), \quad (2)$$

where the notation is similar to that of Equation 1. Here, we again use the output of C_g as a regularizer to train $\widehat{C_g}$. In the evaluation phase, to label the unseen examples, we take the average of the outputs of the two classifiers $\widehat{C_d}$ and $\widehat{C_g}$.

Algorithm 1 illustrates our model (VID) in Structured English. On Lines 8 and 9 we derive the document and drug representations from the sets L and U. On Lines 10 and 11 we use the labeled training examples in the two views to train C_d and C_g. On

[2]Again, we use temperature T in the softmax layer to train using the soft-labels.

Algorithm 1 Overview of VID

1: **procedure** VID
2:   **Given:**
3:     L : Set of labeled documents
4:     U : Set of unlabeled documents
5:   **Return:**
6:     Two classifiers $\widehat{C_d}$ and $\widehat{C_g}$
7:   **Execute:**
8:     Derive two sets of representations L_d and L_g from L
9:     Derive two sets of representations U_d and U_g from U
10:     Use L_d to train classifier C_d
11:     Use L_g to train classifier C_g
12:     Use C_d to probabilistically label U_d
13:     Transfer labels of U_d to U_g and use them to train $\widehat{C_g}$
14:     Finetune $\widehat{C_g}$ using Equation 2
15:     Use C_g to probabilistically label U_g
16:     Transfer labels of U_g to U_d and use them to train $\widehat{C_d}$
17:     Finetune $\widehat{C_d}$ using Equation 1
18:     **Return** $\widehat{C_d}$ and $\widehat{C_g}$

Lines 12-14 we train and finetune $\widehat{C_g}$, and on Lines 15-17 we train and finetune $\widehat{C_d}$. Finally, we return $\widehat{C_d}$ and $\widehat{C_g}$. In the next section, we describe our experimental setup.

4 Experimental Setup

We evaluated our model in the largest publicly available ADR dataset, i.e., the SMM4H dataset. This dataset consists of 30,174 tweets. The training set in this dataset consists of 25,616 tweets of which 9.2% are positive. The labels of the test set are not publicly available. The evaluation in the dataset must be done via the CodaLab website. We compare our model with two sets of baselines: 1) a set of baselines that we implemented, 2) the set of baselines that are available on the CodaLab website[3].

Our own baseline models are: **BERT**, the base variant of the pretrained BERT model (Devlin et al., 2019), as published by Google. **BERT-D**, a domain-specific pretrained BERT model. This model is similar to the previous baseline, however, it is further pretrained on 800K unlabeled drug-related tweets that we collected from Twitter. We pretrained this model for 6 epochs using the next sentence prediction and the masked language model tasks. **BERT-D-BL**, a bi-directional LSTM model. In this model we used BERT-D followed by a bi-directional LSTM network (Hochreiter and Schmidhuber, 1997).

[3]Available at: https://competitions.codalab.org/SMM4H. The 2020 edition of the shared task is not online anymore. Therefore, for a fair comparison with the baselines, we do not use RoBERTa in our model, and instead use pre-trained BERT model.

Type	Method	F1	Precision	Recall
Our Impl.	BERT	0.57	0.669	0.50
	BERT-D	0.62	0.736	0.54
	BERT-D-BL	0.61	**0.749**	0.52
CodaLab	Sarthak	0.65	0.661	0.65
	leebean337	0.67	0.600	**0.76**
	aab213	0.67	0.608	0.75
	VID	**0.70**	0.678	0.72

Table 1: F1, Precision, and Recall of our model (VID) in comparison with the baselines.

We also compare our model with all the baselines available on the CodaLab webpage. These baselines include published and unpublished models. They also cover models that purely rely on machine learning models and those that heavily employ medical resources; see Weissenbacher and Gonzalez-Hernandez (2019) for the summary of a subset of these models.

We used the Pytorch implementation of BERT (Wolf et al., 2019). we used two instances of BERT-D as the classifiers in our model–see Figure 1. Please note that using domain-specific pretrained BERT in our framework makes any improvement very difficult, because the improvement in the performance should be additive. We used the training set of the dataset to tune for our two hyper-parameters T and λ. The optimal values of these two hyper-parameters are 2 and 0.5 respectively. We trained all the models for 5 epochs[4]. During the tuning, we observed that the finetuning stage in our model requires much fewer training steps, therefore, we finetuned for only 1 epoch. In our model, we used the same set of unlabeled tweets that we used to pretrain BERT-D. This verifies that, indeed, our model extracts new information that cannot be extracted using the regular language model pretraining. As required by SMM4H we tuned for F1 measure. In the next section, we report the F1, Precision, and Recall metrics.

5 Results and Analysis

Table 1 reports the performance of our model in comparison with the baseline models–only the top three CodaLab baselines are listed here. We see that our model significantly outperforms all the baseline models. We also observe that the performances of our implemented baseline models are lower than that of the CodaLab models. This difference is mainly due to the gap between the size of the unlabeled sets for the language model pretraining in the experiments–ours is 800K, but the top CodaLab model used a corpus of 1.5M examples. This suggests that our model can potentially achieve a better performance if there is a larger unlabeled corpus available.

[4]We used 20% of the training set for validation, and observed that the models overfit if we train more than 5 epochs.

Method	F1	Precision	Recall
Document-View	0.62	0.736	0.54
Drug-View	0.63	0.706	0.570
Combined-View	0.63	0.745	0.543
VID	0.70	0.678	0.72

Table 2: F1, Precision, and Recall of VID in comparison to the performance of the classifiers trained on the document, drug, and combined views.

Method	F1	Precision	Recall
P-Doc-F-Doc	0.69	0.658	0.71
P-Drug-F-Drug	0.68	0.681	0.68
P-Doc-F-Drug	0.70	0.674	0.72
P-Drug-F-Doc	0.69	0.655	0.72
VID	0.70	0.678	0.72

Table 3: Performance of VID in comparison to the performance of the classifiers pretrained on the document or drug pseudo-labels (indicated by P-$\{*\}$) and finetuned on the document or drug training examples (indicated by F-$\{*\}$).

Table 2 reports the performance of VID in comparison to the classifiers trained on the document and drug representations. We also concatenated the two representations and trained a classifier on the resulting feature vector, denoted by *Combined-View*. We see that our model substantially outperforms all three models. Table 3 compares our model with the classifiers with different pretraining and finetuning resources. Again, we see that VID is comparable to the best of these models. We also observe 2 percent absolute improvement by comparing *P-Drug-F-Drug* and *P-Doc-F-Drug*, which signifies the efficacy of View Distillation.

In summary, we evaluated our model in the largest publicly available ADR dataset and compared with the state-of-the-art baseline models that use domain specific language model pretraining. We showed that our model outperforms these models, even though it uses a smaller unlabeled corpus. We also carried out a set of experiments and demonstrated the efficacy of our proposed techniques.

6 Conclusions

In this study we proposed a novel model for extracting adverse drug effects from user generated content. Our model relies on unlabeled data and a novel technique called view distillation. We evaluated our model in the largest publicly available ADR dataset, and showed that it outperforms the existing BERT-based models.

References

Eiji Aramaki, Sachiko Maskawa, and Mizuki Morita. 2011. Twitter catches the flu: Detecting influenza epidemics using twitter. In *Proceedings of the Conference on Empirical Methods in Natural Language Processing*, EMNLP '11, pages 1568–1576, Stroudsburg, PA, USA. Association for Computational Linguistics.

Jacob Devlin, Ming-Wei Chang, Kenton Lee, and Kristina Toutanova. 2019. BERT: Pre-training of deep bidirectional transformers for language understanding. In *Proc of the 2019 NAACL*, pages 4171–4186.

Mei Sheng Duh, Pierre Cremieux, Marc Van Audenrode, Francis Vekeman, Paul Karner, Haimin Zhang, and Paul Greenberg. 2016. Can social media data lead to earlier detection of drug-related adverse events? *Pharmacoepidemiology and Drug Safety*, 25(12):1425–1433.

Su Golder, Gill Norman, and Yoon K Loke. 2015. Systematic review on the prevalence, frequency and comparative value of adverse events data in social media. *British Journal of Clinical Pharmacology*, 80(4):878–888.

Graciela Gonzalez-Hernandez, Ari Z. Klein, Ivan Flores, Davy Weissenbacher, Arjun Magge, Karen O'Connor, Abeed Sarker, Anne-Lyse Minard, Elena Tutubalina, Zulfat Miftahutdinov, and Ilseyar Alimova, editors. 2020. *Proceedings of the Fifth Social Media Mining for Health Applications Workshop & Shared Task*. Association for Computational Linguistics, Barcelona, Spain (Online).

Geoffrey Hinton, Oriol Vinyals, and Jeff Dean. 2015. Distilling the knowledge in a neural network. *arXiv preprint arXiv:1503.02531*.

Sepp Hochreiter and Jürgen Schmidhuber. 1997. Long short-term memory. *Neural Comput.*, 9(8):1735–1780.

Negin Karisani and Payam Karisani. 2020. Mining coronavirus (covid-19) posts in social media. *arXiv preprint arXiv:2004.06778*.

Payam Karisani and Eugene Agichtein. 2018. Did you just have a heart attack?: Towards robust detection of personal health mentions in social media. In *Proc of the 2018 WWW*, pages 137–146.

Payam Karisani, Eugene Agichtein, and Joyce Ho. 2020. Domain-guided task decomposition with self-training for detecting personal events in social media. In *Proceedings of The Web Conference 2020*, WWW '20, page 2411–2420, New York, NY, USA. Association for Computing Machinery.

Payam Karisani and Negin Karisani. 2021. Semi-supervised text classification via self-pretraining. In *Proceedings of the 14th ACM International Conference on Web Search and Data Mining*, WSDM '21, page 40–48. Association for Computing Machinery.

Payam Karisani, Farhad Oroumchian, and Maseud Rahgozar. 2015. Tweet expansion method for filtering task in twitter. In *Experimental IR Meets Multilinguality, Multimodality, and Interaction*, pages 55–64.

Shasha Liao and Ralph Grishman. 2011. Using prediction from sentential scope to build a pseudo co-testing learner for event extraction. In *Proc of 5th IJCNLP*, pages 714–722.

Yinhan Liu, Myle Ott, Naman Goyal, Jingfei Du, Mandar Joshi, Danqi Chen, Omer Levy, Mike Lewis, Luke Zettlemoyer, and Veselin Stoyanov. 2019. Roberta: A robustly optimized bert pretraining approach.

Alexandra Olteanu, Emre Kiciman, and Carlos Castillo. 2018. A critical review of online social data: Biases, methodological pitfalls, and ethical boundaries. In *Proceedings of the Eleventh ACM International Conference on Web Search and Data Mining*, WSDM '18, pages 785–786, New York, NY, USA. ACM.

Gabriel Pereyra, George Tucker, Jan Chorowski, Lukasz Kaiser, and Geoffrey E. Hinton. 2017. Regularizing neural networks by penalizing confident output distributions. In *5th International Conference on Learning Representations, ICLR 2017, Toulon, France, April 24-26, 2017, Workshop Track Proceedings*. OpenReview.net.

Bianca Scarlini, Tommaso Pasini, and Roberto Navigli. 2020. Sensembert: Context-enhanced sense embeddings for multilingual word sense disambiguation. In *The Thirty-Fourth AAAI Conference on Artificial Intelligence, AAAI 2020, EAAI 2020, New York, NY, USA, February 7-12, 2020*, pages 8758–8765. AAAI Press.

Elena Tutubalina and Sergey Nikolenko. 2017. Combination of deep recurrent neural networks and conditional random fields for extracting adverse drug reactions from user reviews. *Journal of healthcare engineering*, 2017.

Ashish Vaswani, Noam Shazeer, Niki Parmar, Jakob Uszkoreit, Llion Jones, Aidan N. Gomez, Lukasz Kaiser, and Illia Polosukhin. 2017. Attention is all you need. In *Advances in Neural Information Processing Systems 30: Annual Conference on Neural Information Processing Systems 2017, December 4-9, 2017, Long Beach, CA, USA*, pages 5998–6008.

Davy Weissenbacher and Graciela Gonzalez-Hernandez, editors. 2019. *Proceedings of the Fourth Social Media Mining for Health Applications (#SMM4H) Workshop & Shared Task*. Association for Computational Linguistics, Florence, Italy.

Thomas Wolf, Lysandre Debut, and et al. 2019. Huggingface's transformers: State-of-the-art natural language processing. *ArXiv*, abs/1910.03771.

Andrew Yates and Nazli Goharian. 2013. Adrtrace: Detecting expected and unexpected adverse drug reactions from user reviews on social media sites. In *Advances in Information Retrieval*, pages 816–819, Berlin, Heidelberg. Springer Berlin Heidelberg.

The ProfNER shared task on automatic recognition of occupation mentions in social media: systems, evaluation, guidelines, embeddings and corpora

Antonio Miranda-Escalada
Barcelona Supercomputing Center (BSC)
Barcelona, Spain

Eulàlia Farré-Maduell
Barcelona Supercomputing Center (BSC)
Barcelona, Spain

Salvador Lima-López
Barcelona Supercomputing Center (BSC)
Barcelona, Spain

Luis Gascó
Barcelona Supercomputing Center (BSC)
Barcelona, Spain

Vicent Briva-Iglesias
SFI Centre for Research Training in Digitally-Enhanced Reality, Dublin City University
Dublin, Ireland

Marvin M. Agüero-Torales
Barcelona Supercomputing Center (BSC)
Barcelona, Spain

Martin Krallinger
Barcelona Supercomputing Center (BSC)
Barcelona, Spain

{antonio.miranda, eulalia.farre, salvador.limalopez, lgasco, martin.krallinger}@bsc.es
vicent.brivaiglesias2@mail.dcu.ie, maguero@correo.ugr.es

Abstract

Detection of occupations in texts is relevant for a range of important application scenarios, like competitive intelligence, sociodemographic analysis, legal NLP or health-related occupational data mining. Despite the importance and heterogeneous data types that mention occupations, text mining efforts to recognize them have been limited. This is due to the lack of clear annotation guidelines and high-quality Gold Standard corpora. Social media data can be regarded as a relevant source of information for real-time monitoring of at-risk occupational groups in the context of pandemics like the COVID-19 one, facilitating intervention strategies for occupations in direct contact with infectious agents or affected by mental health issues. To evaluate current NLP methods and to generate resources, we have organized the ProfNER track at SMM4H 2021, providing ProfNER participants with a Gold Standard corpus of manually annotated tweets (human IAA of 0.919) following annotation guidelines available in Spanish and English, an occupation gazetteer, a machine-translated version of tweets, and FastText embeddings. Out of 35 registered teams, 11 submitted a total of 27 runs. Best-performing participants built systems based on recent NLP technologies (e.g. transformers) and achieved 0.93 F-score in Text Classification and 0.839 in Named Entity Recognition. Corpus: https://doi.org/10.5281/zenodo.4309356

1 Introduction

The number of social media users and content is rapidly growing, with over 700 million tweets posted daily (James, 2019). The Social Media Mining 4 Health (SMM4H) effort (Klein et al., 2020; Magge et al., 2021) attempts to promote the development and evaluation of NLP and text mining resources to extract automatically relevant health-related information from social media data. As social media content is produced directly by users at a global scale, a variety of application scenarios of medical importance have been explored so far, including the use of social media for pharmacovigilance (Nikfarjam et al., 2015), medication adherence, (Belz et al., 2019) or tracking the spread of infectious and viral diseases (Rocklöv et al., 2019; Zadeh et al., 2019).

The Spanish-speaking community on Twitter is large, exceeding 30 million (Tankovska, 2019), which motivated the implementation of text mining efforts for health-related applications, in particular on drug-related effects (Segura-Bedmar et al., 2015, 2014; Ramírez-Cifuentes et al., 2020).

One of the current challenges for health-related social media applications is to generate more actionable knowledge that can drive the design of intervention plans or policies to improve population health. This is particularly true for infectious disease outbreaks like the COVID-19 pandemic,

Proceedings of the Sixth Social Media Mining for Health Workshop 2021, pages 13–20
June 10, 2021.

where certain occupational groups and subpopulations have been at higher risk, due to direct exposure to infected persons, a higher degree of social interaction, work-related travels to high-risk areas or mental-health problems associated to work-induced stress. Early detection and characterization of at-risk professions is critical to design and prioritize preventive and therapeutic measures or even vaccination plans.

To date, occupational text mining has been used in clinical narratives (Dehghan et al., 2016) and to explore injuries in the construction sector (Cheng et al., 2012), mainly in English. However, it has not yet been systematically used in social media and clinical content in Spanish.

To implement a central NLP component for occupational data mining, namely the automatic detection of occupation mentions in social media texts, we have organized the ProfNER (SMM4H 2021) shared task. In this article, occupation mentions are all those elements that indicate the employment information of a person. Within occupations, we have identified three main labels: (i) "profession", occupations that provide a person with a wage or livelihood (e.g., nurse); (ii) "activity", unpaid occupations (e.g., activist); and (iii) "working status" (e.g., retired).

Resources released for the ProfNER track included annotation guidelines in Spanish and English, a consistency analysis to characterize the quality and difficulty of the track, a large collection of manually annotated occupation mentions in tweets, as well as FastText word embeddings generated from a very large social media dataset.

We foresee that the occupation mention recognition systems resulting from this track could serve as a key component for more advanced text mining tools integrating technologies related to opinion mining, sentiment analysis, gender-inequality analysis, hate speech or fake news detection. Moreover, there is also a clear potential for exploring occupation recognition results for safety management, risk behavior detection and social services intervention strategies.

2 Task Description

2.1 Shared Task Goal

ProfNER focuses on the automatic recognition of professions and working status mentions on Twitter posts related to the COVID-19 pandemic in Spanish.

2.2 Subtracks

We have structured the ProfNER track into two independent subtracks, one related to the classification of whether a tweet actually does mention occupations or not, and another subtrack on finding the exact text spans referring to occupations.

This setting was decided due to the high class imbalance in social media data. Indeed, only 23.3% of the Gold Standard tweets contained mentions of occupations. Then, detecting relevant tweets (subtrack A) may help to detect the occupation mentions (subtrack B).

Subtrack A: Tweet binary classification. This subtrack required binary classification of tweets into those that had at least a single mention of occupations and those that did not mention any.

Subtrack B: NER offset detection and classification. Participants must find the beginning and the end of relevant mentions and classify them in the corresponding category.

2.3 Shared Task Setting

The ProfNER shared task was organized in three phases run on CodaLab[1]:

Practice phase. The training and validation subsets of the Gold Standard were released. During this period, participants built their system and assessed their performance in the validation partition.

Evaluation phase. The test and background partitions were released without annotations. Participants had to generate predictions for the test and the background sets, but they were evaluated only on the test set predictions. This prevented manual annotations and assessed whether systems were able to scale to larger data collections. Each team was allowed to submit up to two runs.

Post-evaluation phase. The competition is kept alive on CodaLab. Interested stakeholders can still assess how their systems perform.

2.4 Evaluation: Metrics and Baseline

For Subtrack A, systems have been ranked based on F1-score for the positive class (tweets that contain a mention). For Subtrack B, the primary evaluation metric has been the micro-averaged F1-score. In this second track, a prediction was successful if its span matched completely the Gold Standard annotation and had the same category. Only PROFESION (profession) and SITUACION LABORAL

[1]https://competitions.codalab.org/competitions/28766

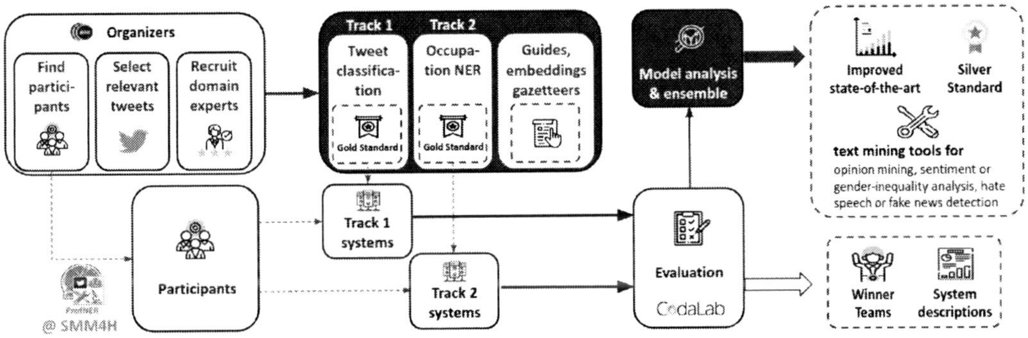

Figure 1: ProfNER Shared Task overview.

(working status) categories are considered in the evaluation.

Also, we have compared every system to a baseline prediction, a Levenshtein lexical lookup approach with a sliding window of varying length.

3 Corpus and Resources

3.1 ProfNER Gold Standard

The ProfNER Gold Standard is a collection of 10,000 COVID-related tweets in Spanish annotated with 4 types of occupation mentions: PROFESION (in English, "profession"), ACTIVIDAD ("activity"), SITUACION LABORAL ("working status") and FIGURATIVA (indicating that occupations are used in a figurative context).

The corpus has been split into training (60%), development (20%) and test (20%) sets. In addition, 25,000 extra tweets are provided without annotations as a background set. Table 1 contains an overview of the corpus statistics.

The corpus was carefully selected to include documents relevant to the challenges of the COVID-19 pandemic. It was obtained from a Twitter crawl that used keywords related to the pandemic (such as "Covid-19") and lockdown measures (like "confinamiento" or "distanciamiento", that are the Spanish translations of lockdown and social distancing), as well as hashtags such as "#yomequedoencasa" (#istayathome), to retrieve relevant tweets. Finally, we only kept the tweets that were written from Spain and in Spanish.

We filtered the tweets using the location attribute of the user profile and looked for the name of Spanish cities with more than 50K inhabitants, province names, autonomous region names, as well as any location specified simply as "Spain".[2]

Gold Standard Quality. The corpus was manually annotated by expert linguists in an iterative process that included the creation of custom-made annotation guidelines, described in Section 3.3. Furthermore, we have performed a consistency analysis of the corpus: 10% of the documents have been annotated by an internal annotator as well as by the expert linguists. The Inter-Annotator Agreement (pairwise agreement) is 0.919.

	Documents	Annotations	Tweets with mentions	Tokens
Training	6,000	1,922	1,393	207,539
Validation	2,000	675	477	68,672
Test	2,000	636	463	68,713
Total	10,000	3,233	2,333	344,924
Background	25,000	81,922 (!)	7,394 (!)	805,961

Table 1: ProfNER corpus summary. (!) Background annotations are extracted from participants' predictions.

Gold Standard Format. Tweets were provided in plain UTF-8 text files, composed Unicode form. Every tweet is contained in a text file whose name is the tweet ID. The tweet classification annotations are distributed in a tab-separated file. The Named Entity Recognition (NER) annotations are distributed in Brat standoff format (Stenetorp et al., 2012) and in a tab-separated file (Fig. 2).

Translation to English. The ProfNER shared task attracted participants from many non-Spanish speaking countries. Besides, there exist more resources for social media text processing in English than in Spanish. For that reason, we have provided, as an additional resource, a machine-translated version of the ProfNER corpus. This will ease the

[2]The list of place names was obtained from the Instituto Nacional de Estadística (INE) `https://www.ine.es/dynt3/inebase/es/index.htm?padre=517&capsel=525`

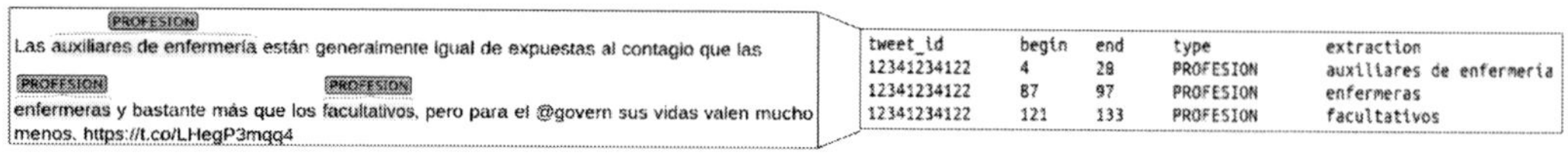

Figure 2: Example of an annotated tweet visualized with Brat annotation tool and its annotation file for track B.

comparison with systems working in English, provide support to participants working previously in English and explore the use of machine-translated corpora.

The complete ProfNER corpus —originally in Spanish— was translated into English by means of a state-of-the-art machine translation system based on recurrent neural networks.

The ProfNER Gold Standard and the English translation are distributed in Zenodo (Miranda-Escalada et al., 2020b).

3.2 ProfNER Silver Standard Corpus

The ProfNER test set was released together with an additional collection of 25,000 tweets as a background set. Participants have generated predictions for the test and background sets. The 81,922 predictions for this background set constitute the ProfNER Silver Standard corpus, similar to the ones of the CALBC (Rebholz-Schuhmann et al., 2010), the CodiEsp (Miranda-Escalada et al., 2020c) and the Cantemist (Miranda-Escalada et al., 2020a) initiatives. They will be released in the Zenodo Medical NLP community[3].

3.3 ProfNER Guidelines

The creation of robust guidelines ensures dataset quality and replicability. Their main objectives are: (1) to capture all possible mentions of the entities of interest, especially occupations (*ex-trazadores de contagios*); and (2) to apply constraints to the mentions in order to obtain well-defined, replicable expressions (*ex-empleada en carpintería mecánica*).

ProfNER's guidelines were created from scratch and iteratively refined to achieve maximum richness of mentions and until the Inter-Annotator Agreement was sufficiently high. All in all, six batches of annotations, corrections and reviews were required, reaching an agreement of 0.919. The final version includes 54 rules that describe the concepts to annotate and the associated constrictions. They are divided in four major groups: (i) 12 general rules, explaining the classification, orthographic and typographical aspects to be considered; (ii) 22 positive rules, explaining what should be deemed as an occupation; (iii) 11 negative rules, showing elements that should not be annotated; and (iv) 9 special cases of annotation. All rules are provided with illustrative corpus examples.

The guidelines were originally written in Spanish and later translated into English by a professional translator; both of them are freely available in Zenodo (Farré-Maduell et al., 2020).

3.4 ProfNER Embeddings

To our knowledge, we have released the first COVID-related embeddings in Spanish. They are especially suited for the ProfNER use case, since they are trained with 140 million COVID-related Spanish Twitter posts.

URL and mentions are substituted by the standard tokens URL and @MENTION, respectively. Embeddings were trained with FastText (Bojanowski et al., 2017) and the chosen embedding size was 300. CBOW and Skip-gram models in cased and uncased versions are available in Zenodo (Miranda-Escalada et al., 2021).

3.5 ProfNER occupations gazetteer

We have released the ProfNER gazetteer of occupations in Spanish, a resource that covers terminological resources from multiple terminologies (DeCS, ESCO, SnomedCT and WordNet) and occupations detected by Stanford CoreNLP in a large collection of social media Spanish profiles. The gazetteer can be found in Zenodo (Asensio et al., 2021).

4 Results

Participation Overview. Since this is the first task on the detection of occupational entities in social media, ProfNER has received considerable attention from the community. Indeed, 35 teams registered for the task (31 for the Subtrack A and 29 for the Subtrack B), and there were 27 submissions (15 in Subtrack A and 12 in Subtrack B) from 11 teams. Participant teams came both from the industry (3) and academia (8), and from different countries such as Spain, Portugal, Romania and China.

[3]https://zenodo.org/communities/medicalnlp/

Team	Country	A/I	Tool	Subtrack A			Subtrack B			Ref
				P	R	F1	P	R	F1	
Recognai	Spain	I	(2020)	0.93	**0.93**	**0.93**	0.84	**0.84**	**0.84**	(2021)
I2C	Spain	A	-	0.92	0.92	0.92	-	-	-	(2021)
Lasige-BioTM	Portugal	A	(2021)	**0.95**	0.89	0.92	0.81	0.66	0.73	(2021)
MIC-NLP	Germany	A&I	-	0.95	0.856	0.9	0.85	0.80	0.82	(2021)
RACAI	Romania	A	-	0.89	0.88	0.89	0.78	0.74	0.76	(2021)
GnaGna	Spain	I	-	0.93	0.78	0.85	-	-	-	-
UoB-NLP	UK	A	-	0.92	0.77	0.83	-	-	-	(2021)
SINAI	Spain	A	-	0.76	0.48	0.59	0.82	0.65	0.73	(2021)
Troy&AbedInTheMorning	Spain	A	(2021)	-	-	-	**0.88**	0.77	0.82	(2021)
CASIA_Unisound	China	A	-	-	-	-	0.86	0.64	0.73	(2021)
Baseline	Spain	A	(2021)	0.75	0.87	0.8	0.36	0.44	0.40	-

Table 2: Best result per team. Best result bolded, second best underlined. A/I stands for Academy/Industry.

System Results. Table 2 shows an overview of participants' results. In the tweet classification subtrack, there are 3 systems with a similar performance (with F1-scores of 0.92 or 0.93), belonging to academia and industry. In the NER subtrack, the best performing system was developed by Recognai, an industry participant, with 0.839 F1-score. This system was based on a transformer architecture. The second best-performing system was from MIC-NLP, a partnership between Siemens AG and the Ludwig Maximilian University of Munich. They obtained a 0.824 F-score combining contextualized embeddings with BiLSTM-CRF. It is noteworthy that the best systems in terms of F1-score were also the systems with the highest recall, but not the highest precision.

Result analysis. Among the entities that ProfNER participants rarely detect, the proportion of SITUACION LABORAL is larger than in the entire corpus. Besides, some mentions with punctuation signs (particularly # or @) and capital letters are rarely predicted. For instance, "ministro del @interiorgob", "OFICIALES DE MESA" or "PENSIONISTAS" are never predicted. Notably, even though correct boundary detection remains a challenge for NER (Li et al., 2020), in our corpus entity length does not seem to influence predictability.

5 Discussion

To the best of our knowledge, ProfNER is the first occupational data mining effort in social media. It is also the first shared task on health applications of social media in Spanish. Specifically for the shared task, we have built a pioneer Gold Standard for Named Entity Recognition (NER) of occupations in social media in Spanish, the ProfNER corpus. It was generated following the ProfNER annotation guidelines that are shared with the community in Spanish and in English. Finally, we have trained and released the first word embeddings for our use case (Twitter posts in Spanish related to COVID-19) and a gazetteer of relevant terms.

In addition, the ProfNER shared task can be used as template for future shared tasks on the recognition of occupations in social media. Indeed, the English translation of the annotation guidelines eases this research possibility.

ProfNER has aroused interest from both academia and industry. Interestingly, teams from non-Spanish speakers have participated in this task. Tweet classification systems reach high performances. However, the detection and classification of occupational data can still be improved.

We propose the use of the whole ecosystem of occupational text mining resources generated in this shared task (corpus, systems, guidelines, etc.) for building and evaluating novel systems that allow subpopulation characterization in social media in the current pandemic. This system has also the potential to assist public health policy makers in the prevention and management of current and future epidemics.

Moreover, beyond classical evaluation scenarios focusing on traditional quality evaluation using metrics like precision and recall, there is also a need to propose shared tasks and community benchmark efforts that do take into account involvement of end users, as was the case of the BioCreative interactive task (Arighi et al., 2011) or the technical evaluation on the robustness and implementation of named entity components (Leitner et al., 2008; Pérez-Pérez et al., 2016), especially when considering the volume and rapid change of social media data.

Acknowledgements

We acknowledge the Encargo of Plan TL (SEDIA) to BSC for funding of the ProfNER Shared Task. We would particularly like to thank the Social Analytics and the High Performance Artificial Intelligence groups of BSC who worked with us to obtain the Twitter corpus. Special mention to Carlos Rodriguez-Penagos, María Jose Rementeria and to Nataly Buslon Valdez. Finally, a big thank you to the rest of the SMM4H organizers for creating the perfect shared task ecosystem.

References

Cecilia N Arighi, Phoebe M Roberts, Shashank Agarwal, Sanmitra Bhattacharya, Gianni Cesareni, Andrew Chatr-Aryamontri, Simon Clematide, Pascale Gaudet, Michelle Gwinn Giglio, Ian Harrow, et al. 2011. Biocreative iii interactive task: an overview. *BMC bioinformatics*, 12(8):1–21.

Alejandro Asensio, Antonio Miranda-Escalada, Marvin M. Agüero-Torales, and Martin Krallinger. 2021. Occupations gazetteer - ProfNER & MEDDOPROF - occupations, professions and working status terms with their associated codes. Funded by the Plan de Impulso de las Tecnologías del Lenguaje (Plan TL).

Anja Belz, Richard Hoile, Elizabeth Ford, and Azam Mullick. 2019. Conceptualisation and annotation of drug nonadherence information for knowledge extraction from patient-generated texts. In *Proceedings of the 5th Workshop on Noisy User-generated Text (W-NUT 2019)*, pages 202–211.

Piotr Bojanowski, Edouard Grave, Armand Joulin, and Tomas Mikolov. 2017. Enriching word vectors with subword information. *Transactions of the Association for Computational Linguistics*, 5:135–146.

Ching-Wu Cheng, Sou-Sen Leu, Ying-Mei Cheng, Tsung-Chih Wu, and Chen-Chung Lin. 2012. Applying data mining techniques to explore factors contributing to occupational injuries in taiwan's construction industry. *Accident Analysis & Prevention*, 48:214–222.

Azad Dehghan, Tom Liptrot, Daniel Tibble, Matthew Barker-Hewitt, and Goran Nenadic. 2016. Identification of occupation mentions in clinical narratives. In *International Conference on Applications of Natural Language to Information Systems*, pages 359–365. Springer.

Eulàlia Farré-Maduell, Salvador Lima López, Vicent Briva-Iglesias, Marvin M. Agüero-Torales, Antonio Miranda-Escalada, and Martin Krallinger. 2020. Profner guidelines. Funded by the Plan de Impulso de las Tecnologías del Lenguaje (Plan TL).

David Fidalgo, Daniel Vila-Suero, Francisco Aranda, and Ignacio Talavera. 2021. System description for profner - smmh: Optimized fine tuning of a pretrained transformer and word vectors. In *Proceedings of the Sixth Social Media Mining for Health Applications Workshop & Shared Task*.

Josh James. 2019. Data never sleeps 7.0. `https://www.domo.com/learn/data-never-sleeps-7`. Accessed: 2021-03-03.

Ari Klein, Ilseyar Alimova, Ivan Flores, Arjun Magge, Zulfat Miftahutdinov, Anne-Lyse Minard, Karen O'connor, Abeed Sarker, Elena Tutubalina, Davy Weissenbacher, et al. 2020. Overview of the fifth social media mining for health applications (# smm4h) shared tasks at coling 2020. In *Proceedings of the Fifth Social Media Mining for Health Applications Workshop & Shared Task*, pages 27–36.

Frances Laureano De Leon, Harish Tayyar Madabushi, and Mark Lee. 2021. UoB at ProfNER 2021: Data Augmentation for Classification Using-Machine Translation. In *Proceedings of the Sixth Social Media Mining for Health Applications Workshop \& Shared Task*.

Florian Leitner, Martin Krallinger, Carlos Rodriguez-Penagos, Jörg Hakenberg, Conrad Plake, Cheng-Ju Kuo, Chun-Nan Hsu, Richard Tzong-Han Tsai, Hsi-Chuan Hung, William W Lau, et al. 2008. Introducing meta-services for biomedical information extraction. *Genome biology*, 9(2):1–11.

Jing Li, Aixin Sun, Jianglei Han, and Chenliang Li. 2020. A survey on deep learning for named entity recognition. *IEEE Transactions on Knowledge and Data Engineering*.

Arjun Magge, Ari Klein, Ivan Flores, Ilseyar Alimova, Mohammed Ali Al-garadi, Antonio Miranda-Escalada, Zulfat Miftahutdinov, Eulàlia Farré-Maduell, Salvador Lima López, Juan M Banda, Karen O'Connor, Abeed Sarker, Elena Tutubalina, Martin Krallinger, Davy Weissenbacher, and Graciela Gonzalez-Hernandez. 2021. Overview of the sixth social media mining for health applications (# smm4h) shared tasks at naacl 2021. In *Proceedings of the Sixth Social Media Mining for Health Applications Workshop & Shared Task*.

J Alberto Mesa Murgado, Ana Belén Parras Portillo, Pilar López-Úbeda, M Teresa Martín-Valdivia, and L Alfonso Urena Lopez. 2021. Identifying professions & occupations in health-related social media using natural language processing. In *Proceedings of the Sixth Social Media Mining for Health Applications Workshop & Shared Task*.

A Miranda-Escalada, E Farré, and M Krallinger. 2020a. Named entity recognition, concept normalization and clinical coding: Overview of the cantemist track for cancer text mining in spanish, corpus, guidelines, methods and results. In *Proceedings of the Iberian Languages Evaluation Forum (IberLEF 2020), CEUR Workshop Proceedings*.

Antonio Miranda-Escalada. 2021. Bsc baseline. https://github.com/tonifuc3m/profner-baseline.

Antonio Miranda-Escalada, Marvin M. Agüero-Torales, and Martin Krallinger. 2021. Spanish covid-19 twitter embeddings in fasttext. Funded by the Plan de Impulso de las Tecnologías del Lenguaje (Plan TL).

Antonio Miranda-Escalada, Vicent Briva-Iglesias, Eulàlia Farré, Salvador Lima López, Marvin M. Agüero-Torales, and Martin Krallinger. 2020b. ProfNER corpus: gold standard annotations for profession detection in Spanish COVID-19 tweets. Funded by the Plan de Impulso de las Tecnologías del Lenguaje (Plan TL).

Antonio Miranda-Escalada, Aitor Gonzalez-Agirre, Jordi Armengol-Estapé, and Martin Krallinger. 2020c. Overview of automatic clinical coding: annotations, guidelines, and solutions for non-english clinical cases at codiesp track of CLEF eHealth 2020. In *Working Notes of Conference and Labs of the Evaluation (CLEF) Forum*, CEUR Workshop Proceedings.

Azadeh Nikfarjam, Abeed Sarker, Karen O'connor, Rachel Ginn, and Graciela Gonzalez. 2015. Pharmacovigilance from social media: mining adverse drug reaction mentions using sequence labeling with word embedding cluster features. *Journal of the American Medical Informatics Association*, 22(3):671–681.

Victoria Pachón Álvarez, Jacinto Mata Vázquez, and Juan Luís Domínguez Olmedo. 2021. Identification of profession & occupation in health-related social media using tweets in spanish. In *Proceedings of the Sixth Social Media Mining for Health Applications Workshop & Shared Task*.

Martin Pérez-Pérez, Gael Pérez-Rodríguez, Obdulia Rabal, Miguel Vazquez, Julen Oyarzabal, Florentino Fdez-Riverola, Alfonso Valencia, Martin Krallinger, and Anália Lourenço. 2016. The markyt visualisation, prediction and benchmark platform for chemical and gene entity recognition at biocreative/chemdner challenge. *Database*, 2016.

Vasile Păis and Maria Mitrofan. 2021. Assessing multiple word embeddings for named entity recognition of professions and occupations in health-related social media. In *Proceedings of the Sixth Social Media Mining for Health Applications Workshop & Shared Task*.

Diana Ramírez-Cifuentes, Ana Freire, Ricardo Baeza-Yates, Joaquim Puntí, Pilar Medina-Bravo, Diego Alejandro Velazquez, Josep Maria Gonfaus, and Jordi Gonzàlez. 2020. Detection of suicidal ideation on social media: multimodal, relational, and behavioral analysis. *Journal of medical internet research*, 22(7):e17758.

Dietrich Rebholz-Schuhmann, Antonio José Jimeno Yepes, Erik M Van Mulligen, Ning Kang, Jan Kors, David Milward, Peter Corbett, Ekaterina Buyko, Elena Beisswanger, and Udo Hahn. 2010. Calbc silver standard corpus. *Journal of bioinformatics and computational biology*, 8(01):163–179.

Recognai. 2020. Recognai system code. https://github.com/recognai/biome-text.

Joacim Rocklöv, Yesim Tozan, Aditya Ramadona, Maquines O Sewe, Bertrand Sudre, Jon Garrido, Chiara Bellegarde de Saint Lary, Wolfgang Lohr, and Jan C Semenza. 2019. Using big data to monitor the introduction and spread of chikungunya, europe, 2017. *Emerging infectious diseases*, 25(6):1041.

Pedro Ruas. 2021. Lasige-biotm system code. https://github.com/lasigeBioTM/LASIGE-participation-in-ProfNER.

Pedro Ruas, Vitor D. T. Andrade, and Francisco M. Couto. 2021. Lasige-biotm at profner: Bilstm-crf and contextual spanish embeddings for named entity recognition and tweet binary classification. In *Proceedings of the Sixth Social Media Mining for Health Applications Workshop & Shared Task*.

Sergio Santamaría Carrasco and Roberto Cuervo Rosillo. 2021. Word embeddings, cosine similarity and deep learning for identification of professions & occupations in health-related social media. In *Proceedings of the Sixth Social Media Mining for Health Applications Workshop & Shared Task*.

Sergio Santamaría. 2021. Troy&abendinthemorning system code. https://github.com/ssantamaria94/ProfNER-SMM4H.

Isabel Segura-Bedmar, Santiago de la Peña González, and Paloma Martínez. 2014. Extracting drug indications and adverse drug reactions from spanish health social media. In *Proceedings of BioNLP 2014*, pages 98–106.

Isabel Segura-Bedmar, Paloma Martínez, Ricardo Revert, and Julián Moreno-Schneider. 2015. Exploring spanish health social media for detecting drug effects. In *BMC medical informatics and decision making*, volume 15, pages 1–9. BioMed Central.

Pontus Stenetorp, Sampo Pyysalo, Goran Topić, Tomoko Ohta, Sophia Ananiadou, and Jun'ichi Tsujii. 2012. Brat: a web-based tool for nlp-assisted text annotation. In *Proceedings of the Demonstrations at the 13th Conference of the European Chapter of the Association for Computational Linguistics*, pages 102–107.

H. Tankovska. 2019. Countries with the most twitter users 2021. https://www.statista.com/statistics/242606/number-of-active-twitter-users-in-selected-countries. Accessed: 2021-03-03.

Usama Yaseen and Stefan Langer. 2021. Neural text classification and stacked heterogeneous embeddings for named entity recognition in smm4h 2021. In *Proceedings of the Sixth Social Media Mining for Health Applications Workshop & Shared Task*.

Amir Hassan Zadeh, Hamed M Zolbanin, Ramesh Sharda, and Dursun Delen. 2019. Social media for nowcasting flu activity: Spatio-temporal big data analysis. *Information Systems Frontiers*, 21(4):743–760.

Tong Zhou, Zhucong Li, Zhen Gan, Baoli Zhang, Yubo Chen, Kun Niu, Jing Wan, Kang Liu, Jun Zhao, Yafei Shi, Weifeng Chong, and Shengping Liu. 2021. Classification, extraction, and normalization : Casia_unisound team at the social media mining for health 2021 shared tasks. In *Proceedings of the Sixth Social Media Mining for Health Applications Workshop & Shared Task*.

Overview of the Sixth Social Media Mining for Health Applications (#SMM4H) Shared Tasks at NAACL 2021

Arjun Magge
University of Pennsylvania
Philadelphia, PA, USA

Ari Z. Klein
University of Pennsylvania
Philadelphia, PA, USA

Antonio Miranda-Escalada
Barcelona Supercomputing Center
Barcelona, Spain

Mohammed Ali Al-garadi
Emory University
Atlanta, GA, USA

Ilseyar Alimova
Kazan Federal University
Kazan, Russia

Zulfat Miftahutdinov
Kazan Federal University
Kazan, Russia

Eulàlia Farré-Maduell
Barcelona Supercomputing Center
Barcelona, Spain

Salvador Lima López
Barcelona Supercomputing Center
Barcelona, Spain

Ivan Flores
University of Pennsylvania
Philadelphia, PA, USA

Karen O'Connor
University of Pennsylvania
Philadelphia, PA, USA

Davy Weissenbacher
University of Pennsylvania
Philadelphia, PA, USA

Elena Tutubalina
Kazan Federal University
Kazan, Russia

Abeed Sarker
Emory University
Atlanta, GA, USA

Juan M. Banda
Georgia State University
Atlanta, Georgia

Martin Krallinger
Barcelona Supercomputing Center
Barcelona, Spain

Graciela Gonzalez-Hernandez
University of Pennsylvania
Philadelphia, PA, USA

{arjun.magge, ariklein, ivan.flores, karoc, dweissen, gragon}@pennmedicine.upenn.edu
{antonio.miranda, eulalia.farre, salvador.limalopez, martin.krallinger}@bsc.es
{alimovailseyar, zulfatmi, tutubalinaev}@gmail.com
{m.a.al-garadi, abeed}@dbmi.emory.edu
jbanda@gsu.edu

Abstract

The global growth of social media usage over the past decade has opened research avenues for mining health related information that can ultimately be used to improve public health. The Social Media Mining for Health Research and Applications (#SMM4H) shared tasks in its sixth iteration sought to advance the use of social media texts such as Twitter for pharmacovigilance, disease tracking and patient centered outcomes. #SMM4H 2021 hosted a total of eight tasks that included reruns of adverse drug effect extraction in English and Russian and newer tasks such as detecting medication non-adherence from Twitter and WebMD forum, detecting self-reported adverse pregnancy outcomes, detecting cases and symptoms of COVID-19, identifying occupations mentioned in Spanish by Twitter users, and detecting self-reported breast cancer diagnosis. The eight tasks included a total of 12 individual subtasks spanning three languages requiring methods for binary classification, multi-class classification, named entity recognition (NER) and entity normalization. With a total of 97 registering teams and 40 teams submitting predictions, the interest in the shared tasks grew by 70% and participation grew by 38% compared to the previous iteration.

1 Introduction

The Social Media Mining for Health (#SMM4H) shared tasks aim to foster community participation in tackling natural language processing (NLP) challenges in social media texts for health applications. The tasks hosted annually attract newer methods for extraction of meaningful health related infor-

Proceedings of the Sixth Social Media Mining for Health Workshop 2021, pages 21–32
June 10, 2021.

mation from noisy social media sources such as Twitter and WebMD where the information of interest is often sparse and noisy. The NLP methods required for the eight tasks spanned the categories of text classification, named entity recognition and entity normalization. Systems developed for the tasks often require the use of NLP techniques such as noise removal, class weighting, undersampling, oversampling, multi-task learning, transfer learning and semi-supervised learning to improve over traditional methods.

The sixth iteration of #SMM4H hosted eight tasks with a total of twelve individual subtasks. Similar to previous years, the most tasks centered around pharmacovigilance (i.e. ADE extraction, medication adherence) and patient centered outcomes (i.e. adverse pregnancy outcomes, breast cancer diagnosis). This year the shared tasks featured the addition of COVID-19 related tasks such as detection of self reported cases of COVID-19 and symptoms of COVID-19, as well as extraction of professions and occupations for the purposes of risk analysis. The individual tasks are listed below:

1. Classification, extraction and normalization of adverse drug effect (ADE) mentions in English tweets
   (a) Classification of tweets containing ADEs
   (b) Span extraction of ADE mentions
   (c) Span extraction and normalization of ADE mentions
2. Classification of Russian tweets for detecting presence of ADE mentions
3. Classification of change in medications regimen on
   (a) Twitter
   (b) WebMD
4. Classification of tweets self-reporting adverse pregnancy outcomes
5. Classification of tweets self-reporting potential cases of COVID-19
6. Classification of COVID-19 tweets containing symptoms
7. Identification of professions and occupations (ProfNER) in Spanish tweets
   (a) Classification of tweets containing mentions of professions and occupations
   (b) Span extraction of professions and occupations
8. Classification of self-reported breast cancer posts on Twitter

Teams interested in participating were allowed to register for one or more tasks/subtasks. On successful registration, teams were provided with annotated training and validation sets of tweets for each task. In total, 97 teams registered for one or more tasks. The annotated datasets contained examples of input text and output labels which the participants could use to train their methods. During the final evaluation period which lasted four days for each task, teams were provided with a evaluation datasets which contained only the input texts. Participants were required to submit label predictions for the input texts which would be evaluated against the annotated labels. The submissions were facilitated through Codalab [1] and participants were allowed to make up to two prediction submissions for each of the subtasks. Of the 97 registered teams, 40 teams submitted one or more predictions towards the shared tasks.

The remainder of the document is as follows, in Section 2, we briefly describe the individual task objectives and research challenges associated with them. In Section 3, we present the evaluation results and a brief summary of each team's best-performing system for each subtask. Appendix A provides the system description papers corresponding to the team numbers.

2 Tasks

2.1 Task 1: Classification, extraction and normalization of ADE mentions in English tweets

The objectives of Task 1 was to develop automated methods to extract adverse drug effects from tweets containing drug mentions for social media pharmacovigilance. Task 1 and their subtasks have been the longest running tasks at SMM4H. This task presented three challenges listed as subtasks in increasing order of complexity wherein in the systems developed must contain one or more components to accomplish the following: (Task 1a) Classify tweets that contain one or more adverse effects (AE) or also known as adverse drug effect (ADE), (Task 1b) Classify the tweets containing ADEs from Task 1a and further extract the text span of reported ADEs in tweets, and (Task 1c) Classify the tweets containing ADE, extract the text span and further normalize these colloquial mentions to their standard concept IDs in the MedDRA ontology's preferred terms.

[1] https://competitions.codalab.org/

The training dataset contains a total of 18,300 tweets with 17,385 tweets for training, 915 tweets for validation (Magge et al., 2020). Participants were allowed to use both training and validation set for training their models for the evaluation stage. The evaluation was performed on 10,984 tweets. The tweets were manually annotated at three levels corresponding to the three subtasks: (a) tweets that contained one or more mentions of ADE had the *ADE* label assigned to them, (b) each ADE was annotated with the starting and ending indices of the ADE mention in the text, and (c) each ADE also contained the normalized MedDRA lower-level term (LLT) that were evaluated at the higher preferred term (PT) level. There are more than 79,000 MedDRA LLT terms and more than 23,000 preferred terms in the MedDRA ontology. The combined test and training dataset contains 2,765 ADE annotations with 669 unique LLT identifiers. The test set contained 257 LLT terms that were not part of the training set, making it important for the developed system to be capable of extracting ADEs that were not part of the training set. While subtasks 1a and 1b presented a class imbalance problem wherein the classification task needs to take into account that only around 7% of the tweets contain ADEs, subtask 1c presented a challenge with the large potential label space. Systems were evaluated and ranked based on the F_1-score for the *ADE* class, overlapping ADE mentions and overlapping ADE mentions with matching PT ids for subtasks 1a, 1b and 1c respectively.

2.2 Task 2: Classification of Russian tweets for detecting presence of ADE mentions

Task 2 presented a similar challenge to Task 1a wherein the designed system is capable of identifying tweets in Russian that contain one or more adverse drug effects. The dataset contains 11,610 tweets for training and validation, with 1073 (9.24%) tweets that report an ADE. The test set contains 9095 tweets, with 778 (8.55%) tweets that report an ADE. All of the Russian tweets were dual annotated; first, three *Yandex.Toloka*[2] annotators' crowd-sourced labels were aggregated into a single label (Dawid and Skene, 1979), and then the tweets were labeled by a second annotator from KFU. Inter-annotator agreement was 0.74 (Cohen's kappa). Systems were evaluated based on the F_1-score for the "positive" class (i.e., tweets that report an adverse effect).

2.3 Task 3: Classification of change in medications regimen in tweets

Task 3 is a binary classification task that involves distinguishing social media posts where users self-declare changing their medication treatments, regardless of being advised by a health care professional to do so. Posts with self-declaration of changes are annotated as "1", other posts are annotated as "0". Such changes are, for example, not filling a prescription, stopping a treatment, changing a dosage, forgetting to take the drugs, etc. This task is the first step toward detecting patients non-adherent to their treatments and their reasons on social media. The data consists of two corpora: 9,830 tweets from Twitter and 12,972 drug reviews from WebMD. Positive and negative tweets are naturally imbalanced with a 10.38 Imbalance Ratio whereas negative and positive WebMD reviews are naturally balanced with a 0.80 Imbalance Ratio. Each corpus is split into a training (5,898 Tweets / 10,378 Reviews), a validation (1,572 Tweets / 1,297 Reviews), and a test subset (2,360 Tweets / 1,297 Reviews). We provided to the participants the training and validation subsets for both corpora and we evaluated on both test subsets independently. We added in the test sets additional reviews and tweets as decoys to avoid manual corrections of the predicted labels. We evaluated participants' systems based on the F_1-score for the "positive" class (i.e., tweets or reviews mentioning a change in medication treatments).

2.4 Task 4: Classification of tweets self-reporting adverse pregnancy outcomes

Despite the prevalence of miscarriage, stillbirth, preterm birth, and low birthweight, their causes remain largely unknown. To enable the use of Twitter data as a complementary resource for epidemiology of these adverse pregnancy outcomes, Task 4 is a binary classification task that involves automatically distinguishing tweets that potentially report a personal experience of an adverse pregnancy outcome ("outcome" tweets) from those that do not ("non-outcome" tweets). The training set (Klein and Gonzalez-Hernandez, 2020) contains 6487 annotated tweets: 3653 (45%) "outcome" tweets (annotated as "1") and 4456 (55%) "non-outcome" tweets (annotated as "0"). The test set contains 1622 annotated tweets: 731 (45%) "outcome" tweets and

[2] https://toloka.yandex.ru/

891 (55%) "non-outcome" tweets. Inter-annotator agreement (Cohen's kappa) was 0.90. Systems were evaluated based on the F_1-score for the "outcome" class.

2.5 Task 5: Classification of tweets self-reporting potential cases of COVID-19

The COVID-19 pandemic has presented challenges for actively monitoring its spread based on testing alone. Task 5 is a binary classification task that involves automatically distinguishing tweets that self-report potential cases of COVID-19 ("potential case" tweets) from those that do not ("other" tweets), where "potential case" tweets broadly include those indicating that the user or a member of the user's household was denied testing for COVID-19, showing symptoms of COVID-19, potentially exposed to cases of COVID-19, or had had experiences that pose a higher risk of exposure to COVID-19. The training set (Klein et al., 2021) contains 7181 tweets: 1148 (16%) "potential case" tweets (annotated as "1") and 6033 (84%) "other" tweets (annotated as "0"). The test set contains 1795 annotated tweets: 308 (17%) "potential case" tweets and 1487 (83%) "other" tweets. Inter-annotator agreement (Cohen's kappa) was 0.77. Systems were evaluated based on the F_1-score for the "potential case" class.

2.6 Task 6: Classification of COVID-19 tweets containing symptoms

Identifying personal mentions of COVID-19 symptoms requires distinguishing personal mentions from other mentions such as symptoms reported by others and references to news articles or other sources. The classification medical symptoms from COVID-19 Twitter posts presents two key issues: First, there is plenty of discourse around news and scientific articles that describe medical symptoms. While this discourse is not related to any user in particular, it enhances the difficulty of identifying valuable user-reported information. Second, many users describe symptoms that other people experience, instead of their own, as they are usually caregivers or relatives of people presenting the symptoms. This makes the task of separating what the user is self-reporting particularly tricky, as the discourse is not only around personal experiences. Task 6 is considered a three-way classification task where the target classes are: (1) self-reports, (2) non-personal reports, and (3) literature/news mentions. In this task, the tweets were sampled from the collections created by Banda et al. (2020b). The sampled tweets were manually annotated by clinicians for extracting long-term patient-reported symptoms of COVID-19 Banda et al. (2020a). The annotated dataset contained a total of 16,067 tweets, 9567 of which were used for training and 6500 used for testing. Systems were evaluated and ranked based on micro-F_1-scores.

2.7 Task 7: Identification of professions and occupations in Spanish tweets (ProfNER)

Extraction of occupations from health-related content is critical for planning public health measures and epidemiological surveillance systems not only in the context of infectious disease outbreaks like COVID-19. Here, occupations refer to paid (profession) and unpaid (activity) working activities, as well as working status such as "student" or "retired". Occupational risks due to exposure to infectious/hazardous agents or mental health conditions linked to occupational stress require systematic extraction of professions from different types of content including user generate contents like social media. Task 7 focused on the detection of occupations from COVID-related tweets in Spanish (the ProfNER corpus[3]). The aim was to enable detection of health-related issues linked to occupations, with special emphasis on the COVID-19 pandemic. In subtask 7a (text classification), participants had to classify tweets containing occupation mentions in Spanish COVID-related tweets and in subtask 7b (named entity recognition), required extraction of text spans mentioning occupations.

This task presents multiple challenges. The classification task had to cope with class imbalance issues, as only 23.3% of the provided tweets mentioned occupations. Secondly, the occupation mention detection required advanced named entity recognition approaches to deal with the heterogeneity and colloquial ways people were referring to occupations in social media. In both subtasks, participating systems had to process noisy user-generated text in Spanish and scale up to a large number of records. For subtask 7a, systems were evaluated and ranked based on the F_1-score for the positive class i.e. tweets containing an occupation mention and for subtask 7b, F_1-score for the *PROFESSION* and *SITUACION_LABORAL* classes where the spans overlap entirely was used.

[3]https://doi.org/10.5281/zenodo.4309356

2.8 Task 8: Classification of self-reported breast cancer posts on Twitter

Breast cancer patients often discontinue their long-term treatments, such as hormone therapy, increasing the risk of cancer recurrence. These discontinuations may be caused by adverse patient-centered outcomes (PCOs) due to hormonal drug side effects or other factors. PCOs are not detectable through laboratory tests and are sparsely documented in electronic health records. Thus, there is a need to explore complementary sources of information for PCOs associated with breast cancer treatments. Social media is a promising resource but extracting true PCOs from it requires the accurate detection of self-reported breast cancer patients. Task 8 focused on developing systems for this first step i.e. identifying tweets with self-reported breast cancer diagnosis. The dataset for Task 8 contained a total of 3815 tweets for training and 1204 tweets for testing. In this task, only about 26% of the tweets contains such self-reports (S) and 74% of the tweets are non-relevant (NR). Systems designed for this task need to automatically identify tweets in the self-reports category. Systems were evaluated based on the F_1-score for the self-reports (S) class.

3 Results

3.1 Task 1: Classification, extraction and normalization of ADE mentions in English tweets

Table 1 presents the results from Task 1. The best performance achieved in task 1a was an F_1-score of 0.61 which initially appears to be 3 percentage points (p.p) lower than previous year's score of 0.64. However on closer examination we find that in addition to the datasets being different, participants in SMM4H 2020 used additional corpora to train their systems. The best performance in ADE extraction i.e. task 1b was an F_1-score of 0.51 which used multi-task learning methods to optimize their models across classification and the NER task. For both tasks 1a and 1b, we find that the systems with the best *Recall* scores ranked the best among all submissions emphasizing the importance of developing systems that account for the class imbalance. The best performance for the overall task of ADE extraction and normalization i.e. task 1c was 0.29 which was achieved by leveraging annotations from other datasets and incorporating semi-supervised learning across corpora similar to previous year's leading system (Miftahutdinov et al., 2020). Overall, the percentage of teams using transformer architectures for subtasks 1a and 1c rose from 80% in SMM4H 2020 to 100% in SMM4H 2021.

3.2 Task 2: Classification of Russian tweets for detecting presence of ADE mentions

In total, 30 teams were registered and 3 teams submitted models' predictions during the evaluation period. Table 2 presents the F_1-score, precision and recall for the ADE class, for each of the teams' best-performing systems and two baselines for Task 2. Compared to last year's results for this task, arithmetic median of all submissions made by teams increased from 0.42 F_1 to 0.51 F_1. Two best-performing systems for this task in #SMM4H 2020 (Klein et al., 2020a) achieved an F_1-score of 0.51 (Gusev et al., 2020; Miftahutdinov et al., 2020), while the best-performing system in #SMM4H 2021 achieved an F_1-score of 0.57. All teams used a transformer-based architecture.

3.3 Task 3: Classification of change in medications regimen in tweets

Despite the interest for task 3, with 29 teams registered, only one team submitted their predictions during the evaluation period. We reported the performances achieved by the best baseline classifiers and the best team's classifiers in Table 3. The leading team chose a standard architecture for their classifier: a transformer encoder followed by an average pooling layer, a linear layer, and a softmax layer for the prediction. They focused on the impact of the corpora used to pre-train two transformers models, BERT and RoBERTa. They evaluated single and ensemble models pre-trained on corpora of different genres and domains - tweets, clinical notes/biomedical research articles, or Wikipedia. While the ensemble of transformers did not improve on the performance of the default BERT-base model used by the baseline on the WebMD corpus, it proves to be beneficial on the imbalanced Twitter corpus. The baseline classifier handles the imbalance of the Twitter corpus by pre-training with active learning a CNN on the WebMD corpus to transfer the knowledge learned on this balanced corpus(Weissenbacher et al., 2020). The team used more successfully a conventional approach by oversampling the positive tweets of the training set and the ensemble of the predictions of several transformer models. These strategies are not exclusive

Task	Team	F_1	P	R	System Summary
Task 1a	4	**0.61**	0.515	**0.752**	RoBERTa undersampling and oversampling
	23	**0.61**	0.552	0.681	RoBERTa + ChemBERTa
	12	0.54	**0.603**	0.489	BERT variants with oversampling and ensemble
	16	0.49	0.592	0.417	BERTweet with automatically curated (pseudo) data
	10	0.46	0.472	0.456	BERT trained with class weights
	20	0.46	0.523	0.409	BERTweet with class weights
	26	0.44	0.491	0.393	Multi-task learning model with BioBERT and class weights
	27	0.40	0.405	0.401	RoBERTa with SMOTE and data augmentation
	22	0.40	0.521	0.327	BERT ensembles with oversampling
	-	0.31	0.500	0.222	-
	24	0.23	0.135	0.726	BERT
Task 1b	26	**0.51**	0.514	**0.514**	Multi-task learning with selective oversampling
	10	0.50	0.555	0.459	RoBERTa with FastText and byte-pair embeddings
	4	0.50	0.493	0.505	RoBERTa
	22	0.49	**0.681**	0.385	-
	16	0.42	0.381	0.475	BERT with BiLSTM+CRF layer
	23	0.40	0.420	0.382	EnDR-BERT with data from CADEC and COMETA corpora
	24	0.37	0.580	0.275	BERT with joint NER and Normalization
Task 1c	23	**0.29**	0.301	0.275	EnDR-BERT with data from CADEC and COMETA corpora
	28	0.24	0.317	0.196	ELMO, CharCNN and Glove in trained jointly
	24	0.24	**0.371**	0.178	BERT with joint NER and Normalization
	16	0.2	0.139	**0.342**	BERTweet and similarity measures
	26	0.16	0.160	0.170	Multi-task learning with selective oversampling

Table 1: Evaluation results for Task 1:Classification, extraction and normalization of ADE mentions in English tweets. Metrics show F_1-scores (F_1), precision (P), and recall (R) for the *ADE* class.

Team	F_1	P	R	System Summary
23	**0.57**	**0.58**	0.57	EnRuDR-BERT + ChemBERTa
-	0.54	0.57	0.52	-
3	0.47	0.39	**0.59**	BERT-based model trained additionally on augmented texts
Baseline #1	0.41	0.40	0.42	CNN-based classifier with FastText embeddings
Baseline #2	0.50	0.45	0.56	BERT-based classifier

Table 2: Evaluation results for Task 2: Classification of Russian tweets for detecting presence of ADE mentions. Metrics show F_1-scores (F_1), precision (P), and recall (R) for the *ADE* class.

Task	Team	F_1	P	R	System Summary
Task 3a	22	**0.68**	**0.72**	**0.64**	Ensemble of 3 transformers-based models trained with oversampling
	Baseline	0.50	0.47	0.53	CNN trained with transfer and active learning
Task 3b	22	0.86	0.84	**0.89**	Ensemble of 4 transformer-based models
	Baseline	**0.87**	**0.87**	0.88	BERT-based

Table 3: Evaluation results for Task 3: Detecting change in medication treatment in tweets (Task 3a) and WebMD reviews (Task 3b). Metrics show F_1-scores (F_1), precision (P), and recall (R) for the *positive* class.

Team	F_1	P	R	System Summary
22	**0.93**	**0.94**	0.92	BERTweet, RoBERTa-Large
15	0.93	0.91	**0.95**	RoBERTa-Large, language model and classifier fine-tuning
27	0.92	0.89	**0.95**	RoBERTa
-	0.78	0.78	0.78	-

Table 4: Evaluation results for Task 4: Classification of tweets self-reporting adverse pregnancy outcomes. Metrics show F_1-scores (F_1), precision (P), and recall (R) for the *positive* class.

to each other and could be used in a common classifier for future work.

3.4 Task 4: Classification of tweets self-reporting adverse pregnancy outcomes

Table 4 presents the precision, recall, and F_1-score for the *outcome* class, for each of the four team's best-performing system for Task 4. The three top-performing systems achieved similar F_1-scores using RoBERTa pre-trained transformer models. The leading team achieved the marginally highest F_1-score (0.93) using an ensemble of RoBERTa and BERTweet pre-trained models. While the leading team also achieved the highest precision (0.94), the highest recall (0.95) was achieved by another team using the RoBERTa model alone. Overall, using a model pre-trained on tweets did not significantly improve performance for this task. The RoBERTa-based classifiers outperformed a BERT-based classifier (F_1-score = 0.88) presented in recent work (Klein et al., 2020b).

3.5 Task 5: Classification of tweets self-reporting potential cases of COVID-19

Table 5 presents the precision, recall, and F_1-score for the "potential case" class, for each of the 14 team's best-performing system for Task 5. The team with the highest performance F_1-score (0.79), precision (0.78), and recall (0.79) used an ensemble of five BERT-based pre-trained transformer models, including models pre-trained on tweets related to COVID-19. To address the class imbalance, the leading team over-sampled the "potential case" class, and further augmented the "potential case" class using paraphrasing via round-trip translation from English into German, and then back into English. Teams placing second and third achieved F_1-scores of 0.77 and 0.76, respectively, using COVID-Twitter-BERT, while the teams (that submitted system descriptions) that achieved F_1-scores of less than 0.76 did not use models pre-trained on tweets related to COVID-19. The leading team outperformed a benchmark classifier presented in recent work (Klein et al., 2021), which was based on COVID-Twitter-BERT and achieved an F_1-score (0.76) similar to that of the teams placing second and third.

3.6 Task 6: Classification of COVID-19 tweets containing symptoms

Table 6 presents the precision, recall, and F_1-score for Task 6. Unsurprisingly 7 out of the top 11 submissions used BERT or variations of it. Some teams fine tuned their models with additional COVID-19 Twitter data. The best performing team used a fine-tuned version of CT-BERT, achieving a 0.95 F1-score. While most models used are more complex deep learning architectures, team 7 managed to score higher than the median submission scores with a less complex multi-layer perceptron classifier. We believe the high scores in this task were due to the somewhat well-balanced dataset provided without the large class imbalance usually seen in Twitter data.

3.7 Task 7: Identification of professions and occupations in Spanish tweets (ProfNER)

Table 7 presents the Tweet Classification (subtask 7b) and Named Entity Recognition results (subtask 7b). In subtasks 7a and 7b, best-performing systems have effectively combined contextual embeddings or language models with the popular architecture RNN-CRF. For instance, the Recognai team has won both subtasks integrating the pre-trained Spanish language model BETO (Cañete et al., 2020) with an RNN-CRF engine built on top of the FastText medical embeddings (Soares et al., 2019). Besides, lighter models have usually been complemented with gazetteers either built from the training data or gathered from popular occupational terminologies.

3.8 Task 8: Classification of self-reported breast cancer posts on Twitter

Table 8 presents the F_1-score, precision and recall for the *self-reports* class (detection of self-reported breast cancer patient) for the participating teams. The leading team achieved a performance of F_1-score of 0.87. The leading team pre-processed the texts by tokenizing and normalizing tokens by replacing URLs with special tokens and replacing emojis with their semantic expressions. The leading team used BERTweet to encode tweet text and make a binary prediction according to the corresponding pooling vector. The analysis of the results shows that almost all top perform teams have achieved similar/comparable precision. However, the best performing team's recall was 5 p.p higher than the other teams which led to overall improve-

Team	F_1	P	R	System Summary
12	**0.79**	**0.78**	**0.79**	BERT ensemble, oversampling, data augmentation
8	0.77	0.77	0.78	COVID-Twitter-BERT
-	0.77	0.77	0.77	-
-	0.76	0.77	0.76	-
-	0.76	0.73	0.78	-
13	0.76	0.78	0.73	COVID-Twitter-BERT, Twitter-RoBERTa-Base, RoBERTa-Large
15	0.75	0.75	0.76	RoBERTa-Large, Task 6 corpus
22	0.75	0.73	0.77	RoBERTa-Base, RoBERTa-Large, BERTweet
25	0.72	0.70	0.73	XLNet, data augmentation
-	0.70	0.74	0.67	-
6	0.67	0.68	0.65	DistillBERT
29	0.53	0.74	0.41	BERT
-	0.43	0.47	0.39	-
-	0.36	0.56	0.26	-

Table 5: Evaluation results for Task 5: Classification of tweets self-reporting potential cases of COVID-19. Metrics show F_1-scores (F_1), precision (P), and recall (R) for the *positive* class.

Team	F_1	P	R	System Summary
8	**0.95**	**0.9477**	**0.9477**	**Fine-Tuned CT-BERT**
22	0.94	0.9449	0.9449	BERT + RoBERTa Large
11	0.94	0.9448	0.9448	XLNet
12	0.94	0.944	0.944	BERT-base ensemble model with data cleaning & domain specific BERT
4	0.94	0.9411	0.9411	BERTweet + 23 million COVID-19 tweets
24	0.94	0.9406	0.9406	Fine-Tuned BERT-base model
7	0.93	0.9337	0.9337	ML Model - multi-layer perceptron classifier
9	0.93	0.9325	0.9325	Fine-Tuned BERT with small-BERT pre-processing
29	0.84	0.8415	0.8415	BiLSTM
6	0.4	0.3951	0.3951	DistilBERT

Table 6: Evaluation results for Task 6: Classification of COVID-19 tweets containing symptoms. Metrics show micro-F_1-scores (F_1), precision (P), and recall (R)

Task	Team	F_1	P	R	System Summary
Task 7a	17	**0.93**	0.93	**0.93**	BETO language model and RNN with pre-trained word vectors
	2	0.92	0.92	0.92	Language model fine-tuned with ProfNER training corpus
	21	0.92	**0.95**	0.89	Data augmentation, BiLSTM-CRF with FLAIR and FastText embeddings
	10	0.90	**0.95**	0.86	Contextualized embeddings with BiLSTM-CRF
	1	0.89	0.89	0.88	NeuroNER with diverse word embeddings and a gazetteer
	-	0.85	0.93	0.78	-
	19	0.83	0.92	0.76	mBERT fine-tuned on augmented data using back-translation
	14	0.59	0.76	0.48	GloVe embeddings, BiLSTM and an occupations dictionary
	Baseline	0.8	0.75	0.87	Levenshtein distance to find mentions from training set
Task 7b	17	**0.84**	0.84	**0.84**	BETO language model and RNN with pre-trained word vectors
	21	0.73	0.81	0.66	Data augmentation, BiLSTM-CRF with FLAIR and FastText embeddings
	10	0.82	0.85	0.79	Contextualized embeddings with BiLSTM-CRF
	1	0.76	0.78	0.74	NeuroNER with diverse word embeddings
	14	0.73	0.82	0.65	CRF with an occupations dictionary
	5	0.82	**0.88**	0.77	Encoder-decoder architecture with attention feeded by different embeddings
	16	0.73	0.86	0.64	Voting of 30 models with Spanish BERT and BiLSTM modules
	Baseline	0.40	0.36	0.44	Levenshtein distance to find mentions from training set

Table 7: Evaluation results for Task 7: Identification of professions and occupations in Spanish tweets (ProfNER). Metrics show F_1-scores (F_1), precision (P), and recall (R) for the *positive* class on task 7a and micro averaged *PROFESSION* and *SITUACION_LABORAL* classes on task7b.

Team	F_1	P	R	System Summary
16	**0.87**	**0.8701**	**0.87**	BERTweet with fast gradient method
18	0.85	**0.8724**	0.8214	BERT-Large, BlueBERT with adversarial fine-tuning
27	0.84	**0.8706**	0.8058	BioBERT with data augmentation
-	0.85	0.86	0.837	-

Table 8: Evaluation results for Task 8: Classification of self-reported breast cancer posts on Twitter. Metrics show F_1-scores (F_1), precision (P), and recall (R) for the *self-reports* class.

ment in F_1-score.

4 Conclusion

This paper presents an overview of the sixth SMM4H shared tasks held in 2021. The shared tasks hosted a total of eight tasks with 12 individual tasks in total. With 40 teams participating in the shared tasks, we find that interest in tasks grew by 38% from the previous year. Analyzing the methods in the submitted systems, we find that the best systems used transformer based models such as BERT and RoBERTa with various techniques for addressing class imbalance. Details of individual systems are available as system description papers cited in Appendix A.

Acknowledgements

The work for #SMM4H 2020 at the University of Pennsylvania was supported by the National Institutes of Health (NIH) National Library of Medicine (NLM) [grant number R01LM011176]. The work on the Russian set of tweets was done at Kazan Federal University and supported by the Russian Science Foundation [grant number 18-11-00284]. The authors would also like to thank Alexis Upshur for her contribution to annotating tweets, Dmitry Ustalov and other members of the *Yandex.Toloka* team for providing credits for the crowd-sourced annotation of Russian tweets, and all those who reviewed system description papers. The work at the Barcelona Supercomputing Center was supported by the Encargo of the Spanish National Plan for the Advancement of Language Technology (PlanTL).

References

Alham Fikri Aji, Made Nindyatama Nityasya, Haryo Akbarianto Wibowo, Radityo Eko Prasojo, and Tirana Fatyanosa. 2021. BERT Goes Brrr: A Venture Towards the Lesser Error in Classifying Medical Self-Reporters on Twitter. In *Proceedings of the Sixth Social Media Mining for Health Applications (#SMM4H) Workshop & Shared Task*, pages 58–64.

Juan M. Banda, Gurdas Viguruji Singh, Osaid H. Alser, and Daniel Prieto-Alhambra. 2020a. Long-term patient-reported symptoms of covid-19: an analysis of social media data. *medRxiv*.

Juan M. Banda, Ramya Tekumalla, Guanyu Wang, Jingyuan Yu, Tuo Liu, Yuning Ding, Katya Artemova, Elena Tutubalina, and Gerardo Chowell. 2020b. A large-scale covid-19 twitter chatter dataset for open scientific research – an international collaboration.

Pavel Blinov. 2021. Text Augmentation Techniques in Drug Adverse Effect Detection Task. In *Proceedings of the Sixth Social Media Mining for Health Applications (#SMM4H) Workshop & Shared Task*, pages 0–0.

Sergio Santamaría Carrasco and Roberto Cuervo Rosillo. 2021. Word Embeddings, Cosine Similarity and Deep Learning for Identification of Professions & Occupations in Health-related Social Media. In *Proceedings of the Sixth Social Media Mining for Health Applications (#SMM4H) Workshop & Shared Task*, pages 74–76.

José Cañete, Gabriel Chaperon, Rodrigo Fuentes, Jou-Hui Ho, Hojin Kang, and Jorge Pérez. 2020. Spanish pre-trained bert model and evaluation data. In *PML4DC at ICLR 2020*.

Joseph Cornelius, Tilia Ellendorff, and Fabio Rinaldi. 2021. Approaching SMM4H with auto-regressive language models and back-translation. In *Proceedings of the Sixth Social Media Mining for Health Applications (#SMM4H) Workshop & Shared Task*, pages 146–148.

Alexander Philip Dawid and Allan M. Skene. 1979. Maximum likelihood estimation of observer error-rates using the EM algorithm. *Journal of the Royal Statistical Society: Series C (Applied Statistics)*, 28(1):20–28.

Frances Adriana Laureano De Leon, Harish Tayyar Madabushi, and Mark Lee. 2021. UoB at ProfNER 2021: Data Augmentation for Classification Using Machine Translation. In *Proceedings of the Sixth Social Media Mining for Health Applications (#SMM4H) Workshop & Shared Task*, pages 115–117.

George-Andrei Dima, Dumitru-Clementin Cercel, and Mihai Dascalu. 2021. Transformer-based Multi-Task Learning for Adverse Effect Mention Analysis in Tweets. In *Proceedings of the Sixth Social Media Mining for Health Applications (#SMM4H) Workshop & Shared Task*, pages 44–51.

Mohab Elkaref and Lamiece Hassan. 2021. A Joint Training Approach to Tweet Classification and Adverse Effect Extraction and Normalization for SMM4H 2021. In *Proceedings of the Sixth Social Media Mining for Health Applications (#SMM4H) Workshop & Shared Task*, pages 91–94.

David Carreto Fidalgo, Daniel Vila-Suero, Francisco Aranda Montes, and Ignacio Talavera Cepeda. 2021. System description for ProfNER - SMM4H: Optimized finetuning of a pretrained transformer and word vectors. In *Proceedings of the Sixth Social Media Mining for Health Applications (#SMM4H) Workshop & Shared Task*, pages 69–73.

Max Fleming, Priyanka Dondeti, Caitlin Dreisbach, and Adam Poliak. 2021. Fine-tuning Transformers for Identifying Self-Reporting Potential Cases and Symptoms of COVID-19 in Tweets. In *Proceedings of the Sixth Social Media Mining for Health Applications (#SMM4H) Workshop & Shared Task*, pages 131–134.

Yuting Guo, Yao Ge, Mohammed Ali Al-Garadi, and Abeed Sarker. 2021. Pre-trained Transformer-based Classification and Span Detection Models for Social Media Health Applications. In *Proceedings of the Sixth Social Media Mining for Health Applications (#SMM4H) Workshop & Shared Task*, pages 52–57.

Andrey Gusev, Anna Kuznetsova, Anna Polyanskaya, and Egor Yatsishin. 2020. Bert implementation for detecting adverse drug effects mentions in russian. In *Proceedings of the Fifth Social Media Mining for Health Applications Workshop & Shared Task*, pages 46–50.

Zongcheng Ji, Tian Xia, and Mei Han. 2021. PAII-NLP at SMM4H 2021: Joint Extraction and Normalization of Adverse Drug Effect Mentions in Tweets. In *Proceedings of the Sixth Social Media Mining for Health Applications (#SMM4H) Workshop & Shared Task*, pages 126–127.

Tanay Kayastha, Pranjal Gupta, and Pushpak Bhattacharyya. 2021. BERT based Adverse Drug Effect Tweet Classification. In *Proceedings of the Sixth Social Media Mining for Health Applications (#SMM4H) Workshop & Shared Task*, pages 88–90.

Ari Z. Klein, Ilseyar Alimova, Ivan Flores, Arjun Magge, Zulfat Miftahutdinov, Anne-Lyse Minard, Karen O'connor, Abeed Sarker, Elena Tutubalina, Davy Weissenbacher, and Graciela Gonzalez-Hernandez. 2020a. Overview of the fifth Social Media Mining for Health Applications (# SMM4H) shared tasks at COLING 2020. In *Proceedings of the Fifth Social Media Mining for Health Applications Workshop & Shared Task*, pages 27–36.

Ari Z. Klein, Haitao Cai, Davy Weissenbacher, Lisa Levine, and Graciela Gonzalez-Hernandez. 2020b. A natural language processing pipeline to advance the use of Twitter data for digital epidemiology of adverse pregnancy outcomes. *Journal of Biomedical Informatics: X*, 8:100076.

Ari Z. Klein and Graciela Gonzalez-Hernandez. 2020. An annotated data set for identifying women reporting adverse pregnancy outcomes on Twitter. *Data in Brief*, 32:106249.

Ari Z. Klein, Arjun Magge, Karen O'Connor, Jesus Ivan Flores Amaro, Davy Weissenbacher, and Graciela Gonzalez-Hernandez. 2021. Towards using Twitter for tracking COVID-19: A natural language processing pipeline and exploratory data set. *Journal of Medical Internet Research*, 23(1):e25314.

Adarsh Kumar, Ojasv Kamal, and Susmita Mazumdar. 2021a. Adversities are all you need: Classification of self-reported breast cancer posts on Twitter using Adversarial Fine-tuning. In *Proceedings of the Sixth Social Media Mining for Health Applications (#SMM4H) Workshop & Shared Task*, pages 112–114.

Deepak Kumar, Nalin Kumar, and Subhankar Mishra. 2021b. NLP@NISER: Classification of COVID19 tweets containing symptoms. In *Proceedings of the Sixth Social Media Mining for Health Applications (#SMM4H) Workshop & Shared Task*, pages 102–104.

Lung-Hao Lee, Man-Chen Hung, Chien-Huan Lu, Chang-Hao Chen, Po-Lei Lee, and Kuo-Kai Shyu. 2021. Classification of Tweets Self-reporting Adverse Pregnancy Outcomes and Potential COVID-19 Cases Using RoBERTa Transformers. In *Proceedings of the Sixth Social Media Mining for Health Applications (#SMM4H) Workshop & Shared Task*, pages 98–101.

Ying Luo, Lis Pereira, and Kobayashi Ichiro. 2021. OCHADAI at SMM4H-2021 Task 5: Classifying self-reporting tweets on potential cases of COVID-19 by ensembling pre-trained language models. In *Proceedings of the Sixth Social Media Mining for Health Applications (#SMM4H) Workshop & Shared Task*, pages 123–125.

Arjun Magge, Elena Tutubalina, Zulfat Miftahutdinov, Ilseyar Alimova, Anne Dirkson, Suzan Verberne, Davy Weissenbacher, and Graciela Gonzalez-Hernandez. 2020. Deepademiner: A deep learning pharmacovigilance pipeline for extraction and normalization of adverse drug effect mentions on twitter. *medRxiv*.

Zulfat Miftahutdinov, Andrey Sakhovskiy, and Elena Tutubalina. 2020. Kfu nlp team at smm4h 2020 tasks: Cross-lingual transfer learning with pretrained language models for drug reactions. In *Proceedings of the Fifth Social Media Mining for Health Applications Workshop & Shared Task*, pages 51–56.

Anupam Mondal, Sainik Mahata, Monalisa Dey, and Dipankar Das. 2021. Classification of COVID19 tweets using Machine Learning Approaches. In *Proceedings of the Sixth Social Media Mining for Health Applications (#SMM4H) Workshop & Shared Task*, pages 135–137.

Alberto Mesa Murgado, Ana Parras Portillo, Pilar Maite Martin López Úbeda, and Alfonso Urena-López. 2021. Identifying professions & occupations in Health-related Social Media using Natural Language Processing. In *Proceedings of the Sixth Social Media Mining for Health Applications (#SMM4H) Workshop & Shared Task*, pages 141–145.

Atul Kr. Ojha, Priya Rani, Koustava Goswami, Bharathi Raja Chakravarthi, and John P. McCrae. 2021. ULD-NUIG at Social Media Mining for

Health Applications (#SMM4H) Shared Task 2021. In *Proceedings of the Sixth Social Media Mining for Health Applications (#SMM4H) Workshop & Shared Task*, pages 149–152.

Victoria Pachón, Jacinto Mata Vázquez, and Juan Luís Domínguez Olmedo. 2021. Identification of profession & occupation in Health-related Social Media using tweets in Spanish. In *Proceedings of the Sixth Social Media Mining for Health Applications (#SMM4H) Workshop & Shared Task*, pages 0–0.

Vasile Pais and Maria Mitrofan. 2021. Assessing multiple word embeddings for named entity recognition of professions and occupations in health-related social media. In *Proceedings of the Sixth Social Media Mining for Health Applications (#SMM4H) Workshop & Shared Task*, pages 0–0.

Varad Pimpalkhute, Prajwal Nakhate, and Tausif Diwan. 2021. Transformers Models for Classification of Health-Related Tweets. In *Proceedings of the Sixth Social Media Mining for Health Applications (#SMM4H) Workshop & Shared Task*, pages 118–122.

Sidharth Ramesh, Abhiraj Tiwari, Parthivi Choubey, Saisha Kashyap, Sahil Khose, Kumud Lakara, Nishesh Singh, and Ujjwal Verma. 2021. BERT based Transformers lead the way in Extraction of Health Information from Social Media. In *Proceedings of the Sixth Social Media Mining for Health Applications (#SMM4H) Workshop & Shared Task*, pages 33–38.

Rajarshi Roychoudhury and Sudip Naskar. 2021. Fine-tuning BERT to classify COVID19 tweets containing symptoms. In *Proceedings of the Sixth Social Media Mining for Health Applications (#SMM4H) Workshop & Shared Task*, pages 138–140.

Pedro Ruas, Vitor Andrade, and Francisco Couto. 2021. Lasige-BioTM at ProfNER: BiLSTM-CRF and contextual Spanish embeddings for Named Entity Recognition and Tweet Binary Classification. In *Proceedings of the Sixth Social Media Mining for Health Applications (#SMM4H) Workshop & Shared Task*, pages 108–111.

Andrey Sakhovskiy, Zulfat Miftahutdinov, and Elena Tutubalina. 2021. KFU NLP Team at SMM4H 2021 Tasks: Cross-lingual and Cross-modal BERT-based Models for Adverse Drug Effects. In *Proceedings of the Sixth Social Media Mining for Health Applications (#SMM4H) Workshop & Shared Task*, pages 39–43.

Felipe Soares, Marta Villegas, Aitor Gonzalez-Agirre, Martin Krallinger, and Jordi Armengol-Estapé. 2019. Medical word embeddings for spanish: Development and evaluation. In *Proceedings of the 2nd Clinical Natural Language Processing Workshop*, pages 124–133.

Alberto Valdes, Jesus Lopez, and Manuel Montes. 2021. UACH at SMM4H: a BERT based approach for classification of COVID-19 Twitter posts. In *Proceedings of the Sixth Social Media Mining for Health Applications (#SMM4H) Workshop & Shared Task*, pages 65–68.

Davy Weissenbacher, Suyu Ge, Ari Klein, Karen O'Connor, Robert Gross, Sean Hennessy, and Graciela Gonzalez-Hernandez. 2020. Active neural networks to detect mentions of changes to medication treatment in social media. *medRxiv*.

Usama Yaseen and Stefan Langer. 2021. Neural Text Classification and Stacked Heterogeneous Embeddings for Named Entity Recognition in SMM4H 2021. In *Proceedings of the Sixth Social Media Mining for Health Applications (#SMM4H) Workshop & Shared Task*, pages 83–87.

Tong Zhou, Zhucong Li, Zhen Gan, Baoli Zhang, Yubo Chen, Kun Niu, Jing Wan, Kang Liu, Jun Zhao, Yafei Shi, Weifeng Chong, and Shengping Liu. 2021. Classification, Extraction, and Normalization : CASIA_Unisound Team at the Social Media Mining for Health 2021 Shared Tasks. In *Proceedings of the Sixth Social Media Mining for Health Applications (#SMM4H) Workshop & Shared Task*, pages 77–82.

Appendix A. Team Numbers and System Description Papers

Team	System Description Paper
1	(Pais and Mitrofan, 2021)
2	(Pachón et al., 2021)
3	(Blinov, 2021)
4	(Ramesh et al., 2021)
5	(Carrasco and Rosillo, 2021)
6	(Fleming et al., 2021)
7	(Mondal et al., 2021)
8	(Valdes et al., 2021)
9	(Roychoudhury and Naskar, 2021)
10	(Yaseen and Langer, 2021)
11	(Kumar et al., 2021b)
12	(Aji et al., 2021)
13	(Luo et al., 2021)
14	(Murgado et al., 2021)
15	(Lee et al., 2021)
16	(Zhou et al., 2021)
17	(Fidalgo et al., 2021)
18	(Kumar et al., 2021a)
19	(De Leon et al., 2021)
20	(Kayastha et al., 2021)
21	(Ruas et al., 2021)
22	(Guo et al., 2021)
23	(Sakhovskiy et al., 2021)
24	(Elkaref and Hassan, 2021)
25	(Cornelius et al., 2021)
26	(Dima et al., 2021)
27	(Pimpalkhute et al., 2021)
28	(Ji et al., 2021)
29	(Ojha et al., 2021)

Table 9: Key for identifying teams in the Results section. Identifier in the parenthesis is the publication id associated with the SMM4H Workshop.

BERT based Transformers lead the way in Extraction of Health Information from Social Media

Sidharth Ramesh[1] **Abhiraj Tiwari**[1] **Parthivi Choubey**[1] **Saisha Kashyap**[2]

Sahil Khose[2] **Kumud Lakara**[1] **Nishesh Singh**[3] **Ujjwal Verma**[4]

{sidram2000, abhirajtiwari, parthivichoubey, saishakashyap8, sahilkhose18, lakara.kumud, singhnisheshh4}@gmail.com
ujjwal.verma@manipal.edu

Abstract

This paper describes our submissions for the Social Media Mining for Health (SMM4H) 2021 shared tasks. We participated in 2 tasks: (1) Classification, extraction and normalization of adverse drug effect (ADE) mentions in English tweets (Task-1) and (2) Classification of COVID-19 tweets containing symptoms (Task-6). Our approach for the first task uses the language representation model RoBERTa with a binary classification head. For the second task, we use BERTweet, based on RoBERTa. Fine-tuning is performed on the pre-trained models for both tasks. The models are placed on top of a custom domain-specific pre-processing pipeline. Our system ranked first among all the submissions for subtask-1(a) with an F1-score of 61%. For subtask-1(b), our system obtained an F1-score of 50% with improvements up to +8% F1 over the score averaged across all submissions. The BERTweet model achieved an F1 score of 94% on SMM4H 2021 Task-6.

1 Introduction

Social media platforms are a feature of everyday life for a large proportion of the population with an estimated 4.2 billion people using some form of social media (Hootsuite and Social, 2021). Twitter is one of the largest social media platforms with 192 million daily active users (Conger, 2021). The 6th Social Media Mining for Health Applications Workshop focuses on the use of Natural Language Processing (NLP) for a wide number of tasks related to Health Informatics using data extracted from Twitter .

Our team, TensorFlu, participated in 2 tasks, (1) Task-1: Classification, extraction and normalization of adverse effect (AE) mentions in English tweets and (2) Task-6: Classification of COVID-19 tweets containing symptoms. A detailed overview of the shared tasks in the 6th edition of the workshop can be found in (Magge et al., 2021).

The classification and extraction of Adverse Drug Effects (ADE) on social media can be a useful indicator to judge the efficacy of medications and drugs while ensuring that any side effects that previously remained unknown can be found. Thus social media can be a useful medium to judge gauge patient satisfaction and well being.

According to the report in (Shearer and Mitchell, 2021), 15% of American adults get their news on Twitter while 59% of Twitter users get their news on Twitter itself. Thus during the spread of a pandemic like COVID-19, tracking reports by users as well as news mentions from local organizations can perform the function of tracking the spread of the disease in new regions and keep people informed.

Similar to the last edition of the workshop, the top performing model (Klein et al., 2020) for Task-1 with the highest score this year was RoBERTa (Liu et al., 2019). The biggest challenge while dealing with the dataset provided for this years competition was the huge class imbalance. The proposed approach handles this by the use of Random Sampling (Abd Elrahman and Abraham, 2013) of the dataset during finetuning. Named Entity Recognition (NER) for the extraction of text spans was performed using the RoBERTa based model provided in the spaCy (Honnibal et al., 2020) `en_core_web_trf` pipeline. For the classification of tweets with COVID-19 symptoms, we used a model called BERTweet (Nguyen et al., 2020) trained on 845 million English tweets and 23 million COVID-19 related English tweets as of the latest publically available version of the model. Fine-tuning was performed on the pretrained models for

[1]Dept. of Computer Science and Engineering
[2]Dept. of Information and Communication Technology
[3]Dept. of Mechanical and Manufacturing Engineering
[4]Dept. of Electronics and Communication Engineering
Manipal Institute of Technology, Manipal Academy of Higher Education, Manipal, India

Proceedings of the Sixth Social Media Mining for Health Workshop 2021, pages 33–38
June 10, 2021.

both tasks. Section 2 summarizes the methodology and results obtained for Task-1, while Section 3 summarizes the methodology and results for Task-6.

2 Task-1: Classification, extraction and normalization of adverse effect (AE) mentions in English tweets

2.1 Sub-Task 1a: Classification

The goal of this sub-task is to classify tweets that contain an adverse effect (AE) or also known as adverse drug effect (ADE) with the label ADE or NoADE.

2.1.1 Data and Pre-processing

The organizers of SMM4H provided us with a training set consisting of 18,256 tweets with 1,297 positive examples and 16,959 negative examples. Thus, the dataset has a huge class imbalance. The validation dataset has 913 tweets with 65 positive examples and 848 negative examples.

To overcome the class imbalance we performed random oversampling and undersampling (Abd Elrahman and Abraham, 2013) on the provided dataset. The dataset was first oversampled using a sampling strategy of 0.1 i.e. the minority class was oversampled so that it was 0.1 times the size of majority class, then the resultant dataset was undersampled using a sampling strategy of 0.5 i.e. the majority class was undersampled so that the majority class was 2 times the size of minority class

Removal of twitter mentions, hashtags and URLs was performed, but it negatively affected the performance of the model. Hence, this pre-processing step was not performed in the final model. The tweets were then preprocessed using fairseq (Ott et al., 2019) preprocessor which tokenizes the sentences using GPT-2 byte pair encoding(Radford et al., 2019) and finally converts them into binary samples.

2.1.2 System Description

Fairseq's (Ott et al., 2019) pretrained RoBERTa (Liu et al., 2019) large model was used for the task with a binary classification head. The RoBERTa model was pretrained over 160GB of data from BookCorpus (Zhu et al., 2015), CC-News (Nagel, 2016), OpenWebText (Gokaslan* et al., 2019) and Stories.

2.1.3 Experiments

RoBERTa and BioBERT (Lee et al., 2019) were trained for ADE classification and extensive hyperparameter tuning was carried out. The hyperparameters tested on the validation split included the learning rate, batch size, and sampling strategy of the dataset. The RoBERTa model was trained for 6 epochs with a batch size of 8. The learning rate was warmed up for 217 steps with a weight decay of 0.1 and a peak learning rate of 10^{-5} for the polynomial learning rate scheduler. A dropout rate of 0.1 is used along with the Adam optimizer having (β_1, β_2)=(0.9, 0.98).

2.1.4 Results

Precision is defined as the ratio between true positives and the sum of true positives and false positives.

$$Precision = \frac{TP}{TP + FP} \quad (1)$$

Recall is defined as the ratio between true positives and the sum of true positives and false negatives.

$$Recall = \frac{TP}{TP + FN} \quad (2)$$

Our primary objective is to create a model that prevents incorrect classification of ADE tweets. A model with higher recall than precision is more desirable for us as the former tends to reduce the total number of false negatives. F1 Score is chosen to be the evaluation metric for all our models.

$$F1\text{-}score = \frac{2 \cdot Precision \cdot Recall}{Precision + Recall} \quad (3)$$

Table 1 showcases the performance of the different models which performed well on the validation set. The RoBERTa model that was finally chosen after hyperparameter tuning achieved the highest score on the leaderboard among all teams participating in the subtask. The score obtained on the test set can be found in Table 2.

It can be seen in the results of the validation set and test for the ADE class that the recall is 0.92 for the validation set and 0.752 for the test set. The results show that the model has learnt features for classifying ADE samples from a small amount of data. Although it might classify some amount of NoADE tweets incorrectly as evidenced by the low precision, the greater number of correctly classified ADE tweets aligns with our objective of classifying the maximum number of ADE tweets correctly as

S.No.	Model	Arch	Label	Precision	Recall	F1
1.	**RoBERTa**	$BERT_{LARGE}$	NoADE	**0.99**	0.95	0.97
			ADE	0.59	**0.92**	**0.72**
2.	BioBERT	$BERT_{BASE}$	NoADE	0.97	**0.99**	**0.98**
			ADE	**0.78**	0.60	0.68

Table 1: Comparing different models used for task 1a on the **Validation Set**. **RoBERTa** is chosen owing to its higher F1- score while predicting the ADE label correctly.

	Precision	Recall	F1
RoBERTa	**0.515**	**0.752**	**0.61**
Median	0.505	0.409	0.44

Table 2: Comparing our best-performing model to the median for task 1a.

possible so that we don't lose valuable information about adverse drug effects that might be found. Our model achieved a significantly higher recall than the median of all other teams (Table 2), indicating that a majority of ADE tweets are correctly classified.

2.2 Task-1b: ADE Span Detection

The goal of this subtask is to detect the text span of reported ADEs in tweets.

2.2.1 Data and Pre-processing

The given dataset consisted of 1,712 spans across 1,234 tweets. For the purpose of better training of the model, all tweets with duplicate or overlapping spans were manually removed. The decision to do this manually was to ensure that spans providing better context were kept instead of just individual words that would have been less helpful in discerning the meaning of the sentence.

2.2.2 System Description

The dataset was passed through a Named Entity Recognition (NER) pipeline made using the `en_core_web_trf` model. The pipeline makes use of the `roberta-base` model provided by Huggingface's Transformers library (Wolf et al., 2020). The algorithm for extracting Adverse Effects from tweets is provided in Algorithm 1.

Algorithm 1: Algorithm for Extraction of Adverse Drug Effects from Tweets

Input: Input raw tweet T;
Output: $Label$, Start char, End char, Span;
Given (T), Classify the tweet with fairseq RoBERTa into ADE or NoADE;
if $Label$ *is ADE* **then**
  Perform NER on T using spaCy NER pipeline;
  Return Start char, End char, Span;
end

2.2.3 Experiments

Two Named Entity Recognition (NER) pipelines, `en_core_web_trf` (`https://spacy.io/models/en#en_core_web_trf`) and `en_core_web_sm` (`https://spacy.io/models/en#en_core_web_sm`) were tried. The first is a RoBERTa based model while the second is a fast statistical entity recognition system trained on written web text that includes vocabulary, vectors, syntax and entities. After hyperparameter tuning, the transformer model was chosen. The model was trained for 150 epochs with a dropout of 0.3, Adam optimizer (Kingma and Ba, 2014) and a learning rate of 0.001 with (β_1, β_2)=(0.9, 0.999).

2.2.4 Results

The models have been evaluated with two metrics, the Relaxed F1 score, and the Strict F1 score. The Relaxed metrics evaluate the scores for spans that have a partial or full overlap with the labels. The Strict metrics only evaluate the cases where the spans produced by the model perfectly match the span in the label.

Table 3 showcases the performance of both NER pipelines on the validation set. It can be observed that the RoBERTa model provides a higher F1 score than the statistical model and is able to make much more accurate classifications of the ADE class. The statistical model however provides a higher recall which indicates it has fewer false negatives and is thus misclassifying the ADE samples as NoADE less often. The RoBERTa model is however far superior to the statistical model when considering the strict F1 scores. This implies that it is able to produce a perfect span more often and has learnt a

Model	Relaxed P	Relaxed R	Relaxed F1	Strict P	Strict R	Strict F1
`en_core_web_sm`	0.516	**0.551**	0.533	0.226	0.241	0.233
`en_core_web_trf`	**0.561**	0.529	**0.544**	**0.275**	**0.253**	**0.263**

Table 3: Scores on the Validation Set for the model for task 1b.

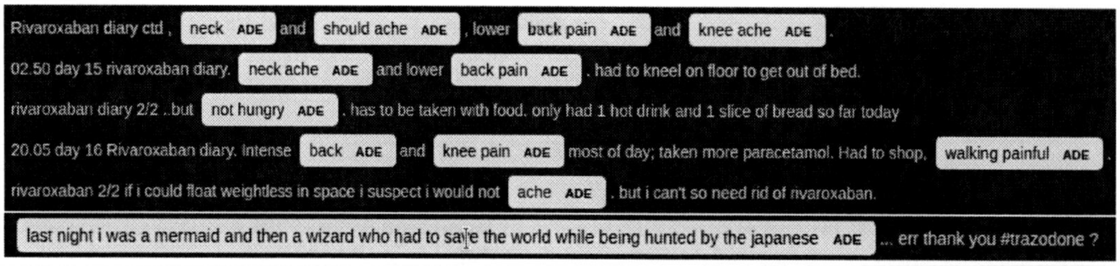

Figure 1: Example span extraction from TensorFlu's model for task 1b

	Precision	Recall	F1
`en_core_web_trf`	0.493	**0.505**	**0.50**
`en_core_web_sm`	**0.521**	0.458	0.49
Median	0.493	0.458	0.42

Table 4: Comparing our best-performing model to the median for task 1b.

better representation of the data.

The final test set result achieved by the model placed on the leaderboard was achieved by the RoBERTa based NER model. The results obtained by both models are compared to the median in Table 4. The transformer pipeline provides a higher recall than the statistical pipeline thus showcasing the fact that a higher number of tweets were correctly classified as ADE while having overlapping spans. A few example images showing the performance of the entire adverse effect extraction pipeline are provided in Figure 1.

3 Task-6: Classification of COVID-19 tweets containing symptoms

The goal of this task is to classify tweets into 3 categories: (1) Self-reports (2) Non-personal reports (3) Literature/News mentions.

3.1 Data and Pre-processing

The SMM4H organizers released a training dataset consisting of 9,567 tweets and test data consisting of 6,500 tweets. The training dataset consisted of 4,523 tweets with Literature/News mentions, 3,622 tweets with non-personal reports and 1,421 tweets with self-reports. There is very little class imbalance in the given dataset. Tokenization of tweets was done using VinAI's `bertweet-base` tokenizer from the *Huggingface* API (Wolf et al., 2020). In order to use the BERTweet model, the tweets were normalized by converting user mentions into the @USER special token and URLs into the HTTPURL special token. The *emoji* package was used to convert the emoticons into text. (Nguyen et al., 2020)

3.2 System Description

BERTweet (Nguyen et al., 2020) uses the same architecture as BERT base and the pre-training procedure is based on RoBERTa, (Liu et al., 2019) for more robust performance, as it optimizes the BERT pre-training approach. BERTweet is optimized using Adam optimizer (Kingma and Ba, 2014), with a batch size of 7K and a peak learning rate of 0.0004, and is pre-trained for 40 epochs (using first 2 epochs for warming up the learning rate). The `bertweet-covid19-base-uncased` model was used for our application, which has 135M parameters, and is trained on 845M English tweets and 23M COVID-19 English tweets.

For training the BERTweet model on our train dataset, (`https://github.com/VinAIResearch/BERTweet`) was used with number of labels set to 3.

3.3 Experiments

A number of experiments were carried out to reach the optimal results for the task. Other models besides BERTweet were trained for the task such as RoBERTa (Liu et al., 2019), DeBERTa (He et al., 2021), and Covid-Twitter-BERT (Müller et al., 2020). A majority voting ensemble with

S.No.	Model	Arch	Label	Precision	Recall	F1
1.	RoBERTa	$BERT_{LARGE}$	Lit-News mentions	0.98	0.97	0.98
			Nonpersonal reports	0.95	0.97	0.96
			Self reports	0.97	0.96	0.97
2.	**BERTweet**	$BERT_{BASE}$	Lit-News mentions	**0.99**	0.99	**0.99**
			Nonpersonal reports	**0.99**	**0.98**	**0.98**
			Self reports	0.97	**1.00**	**0.99**
3.	DeBERTa	$BERT_{BASE}$	Lit-News mentions	0.95	**1.00**	0.98
			Nonpersonal reports	**0.99**	0.95	0.97
			Self reports	**1.00**	0.95	0.97
4.	Covid-Twitter BERT	$BERT_{LARGE}$	Lit-News mentions	0.98	0.98	0.98
			Nonpersonal reports	0.97	0.97	0.97
			Self reports	0.97	0.99	0.98
5.	Majority Voting	NA	Lit-News mentions	0.98	0.99	**0.99**
			Nonpersonal reports	0.98	0.97	0.97
			Self reports	0.99	0.99	**0.99**

Table 5: Comparing different models used for task 6 on the **Validation Set**

all 4 models was also evaluated. After a lot of tuning, BERTweet was found to be the best performing model on the dataset.

The ideal hyperparameters for the model were found empirically following many experiments with the validation set. The best results were obtained with the following hyperparameters: the model was finetuned for 12 epochs with a batch size of 16; the learning rate was warmed up for 500 steps with a weight decay of 0.01.

Due to little class imbalance in the given dataset and pretrained BERT based models performing very well on classification tasks, almost all models achieved a relatively high F1-score.

3.4 Results

The results on the validation set for all the trained models are reported in Table 5. As mentioned in section 2.1.4 the models have been compared on the basis of Precision, Recall and F1-score. The best performing model as seen in Table 5 is BERTweet. The same model was also able to achieve an F1 score above the median on the test set as seen in Table 6.

	Precision	Recall	F1
BERTweet	**0.9411**	**0.9411**	**0.94**
Median	0.93235	0.93235	0.93

Table 6: Comparing our best-performing model to the median for task 6

4 Conclusion

In this work we have explored an application of RoBERTa to the task of classification, extraction and normalization of Adverse Drug Effect (ADE) mentions in English tweets and the application of BERTweet to the task of classification of tweets containing COVID-19 symptoms. We have based our selection of these models on a number of experiments we conducted to evaluate different models. Our experiments have shown that RoBERTa outperforms BioBERT, achieving state of the art results in ADE classification. For the second task, we found that BERTweet outperformed all the other models including an ensembling approach (majority voting).

We foresee multiple directions for future research. One possible improvement could be to use joint learning to deal with Task-1(a) and Task-1(b) simultaneously.

5 Acknowledgements

For Task 1a, *Sidharth R, Abhiraj Tiwari, Parthivi Choubey and Saisha Kashyap* contributed equally to the work. For Task 1b, *Sidharth R, Abhiraj Tiwari, Sahil Khose, and Kumud Lakara* contributed equally to the work. All authors contributed equally to the work done for Task 6. We would also like to thank *Mars Rover Manipal and Project Manas* for providing the necessary resources to train our models.

References

Shaza M Abd Elrahman and Ajith Abraham. 2013. A review of class imbalance problem. *Journal of Network and Innovative Computing*, 1(2013):332–340.

Kate Conger. 2021. Twitter shakes off the cobwebs with new product plans. *The New York Times*.

Aaron Gokaslan*, Vanya Cohen*, Ellie Pavlick, and Stefanie Tellex. 2019. Openwebtext corpus.

Pengcheng He, Xiaodong Liu, Jianfeng Gao, and Weizhu Chen. 2021. Deberta: Decoding-enhanced bert with disentangled attention.

Matthew Honnibal, Ines Montani, Sofie Van Landeghem, and Adriane Boyd. 2020. spaCy: Industrial-strength Natural Language Processing in Python.

Hootsuite and We Are Social. 2021. Digital 2021: Global overview report.

Diederik P. Kingma and Jimmy Ba. 2014. Adam: A method for stochastic optimization. Cite arxiv:1412.6980Comment: Published as a conference paper at the 3rd International Conference for Learning Representations, San Diego, 2015.

Ari Klein, Ilseyar Alimova, Ivan Flores, Arjun Magge, Zulfat Miftahutdinov, Anne-Lyse Minard, Karen O'Connor, Abeed Sarker, Elena Tutubalina, Davy Weissenbacher, and Graciela Gonzalez-Hernandez. 2020. Overview of the fifth social media mining for health applications (#SMM4H) shared tasks at COLING 2020. In *Proceedings of the Fifth Social Media Mining for Health Applications Workshop & Shared Task*, pages 27–36, Barcelona, Spain (Online). Association for Computational Linguistics.

Jinhyuk Lee, Wonjin Yoon, Sungdong Kim, Donghyeon Kim, Sunkyu Kim, Chan Ho So, and Jaewoo Kang. 2019. BioBERT: a pre-trained biomedical language representation model for biomedical text mining. *Bioinformatics*, 36(4):1234–1240.

Yinhan Liu, Myle Ott, Naman Goyal, Jingfei Du, Mandar Joshi, Danqi Chen, Omer Levy, Mike Lewis, Luke Zettlemoyer, and Veselin Stoyanov. 2019. Roberta: A robustly optimized bert pretraining approach.

Arjun Magge, Ari Klein, Ivan Flores, Ilseyar Alimova, Mohammed Ali Al-garadi, Antonio Miranda-Escalada, Zulfat Miftahutdinov, Eulàlia Farré-Maduell, Salvador Lima López, Juan M Banda, Karen O'Connor, Abeed Sarker, Elena Tutubalina, Martin Krallinger, Davy Weissenbacher, and Graciela Gonzalez-Hernandez. 2021. Overview of the sixth social media mining for health applications (# smm4h) shared tasks at naacl 2021. In *Proceedings of the Sixth Social Media Mining for Health Applications Workshop & Shared Task*.

Martin Müller, Marcel Salathé, and Per E Kummervold. 2020. Covid-twitter-bert: A natural language processing model to analyse covid-19 content on twitter.

Sebastian Nagel. 2016. Cc-news.

Dat Quoc Nguyen, Thanh Vu, and Anh Tuan Nguyen. 2020. BERTweet: A pre-trained language model for English tweets. In *Proceedings of the 2020 Conference on Empirical Methods in Natural Language Processing: System Demonstrations*, pages 9–14, Online. Association for Computational Linguistics.

Myle Ott, Sergey Edunov, Alexei Baevski, Angela Fan, Sam Gross, Nathan Ng, David Grangier, and Michael Auli. 2019. fairseq: A fast, extensible toolkit for sequence modeling. In *Proceedings of the 2019 Conference of the North American Chapter of the Association for Computational Linguistics (Demonstrations)*, pages 48–53, Minneapolis, Minnesota. Association for Computational Linguistics.

Alec Radford, Jeff Wu, Rewon Child, David Luan, Dario Amodei, and Ilya Sutskever. 2019. Language models are unsupervised multitask learners.

Elisha Shearer and Amy Mitchell. 2021. News use across social media platforms in 2020. *Pew Research Center*.

Thomas Wolf, Lysandre Debut, Victor Sanh, Julien Chaumond, Clement Delangue, Anthony Moi, Pierric Cistac, Tim Rault, Rémi Louf, Morgan Funtowicz, Joe Davison, Sam Shleifer, Patrick von Platen, Clara Ma, Yacine Jernite, Julien Plu, Canwen Xu, Teven Le Scao, Sylvain Gugger, Mariama Drame, Quentin Lhoest, and Alexander M. Rush. 2020. Huggingface's transformers: State-of-the-art natural language processing.

Yukun Zhu, Ryan Kiros, Rich Zemel, Ruslan Salakhutdinov, Raquel Urtasun, Antonio Torralba, and Sanja Fidler. 2015. Aligning books and movies: Towards story-like visual explanations by watching movies and reading books. In *The IEEE International Conference on Computer Vision (ICCV)*.

KFU NLP Team at SMM4H 2021 Tasks: Cross-lingual and Cross-modal BERT-based Models for Adverse Drug Effects

Andrey Sakhovskiy
Kazan Federal University
Kazan, Russia

Zulfat Miftahutdinov
Kazan Federal University
Kazan, Russia

Elena Tutubalina
Kazan Federal University
Kazan, Russia
HSE University
Moscow, Russia

{andrey.sakhovskiy, zulfatmi, tutubalinaev}@gmail.com

Abstract

This paper describes neural models developed for the Social Media Mining for Health (SMM4H) 2021 Shared Task. We participated in two tasks on classification of tweets that mention an adverse drug effect (ADE) (Tasks 1a & 2) and two tasks on extraction of ADE concepts (Tasks 1b & 1c). For classification, we investigate the impact of joint use of BERT-based language models and drug embeddings obtained by chemical structure BERT-based encoder. The BERT-based multimodal models ranked first and second on classification of Russian (Task 2) and English tweets (Task 1a) with the F1 scores of 57% and 61%, respectively. For Task 1b and 1c, we utilized the previous year's best solution based on the EnDR-BERT model with additional corpora. Our model achieved the best results in Task 1c, obtaining an F1 of 29%.

1 Introduction

Text classification, named entity recognition, and medical concept normalization in free-form texts are crucial steps in every text-mining pipeline. Here we focus on discovering adverse drug effects (ADE) concepts in Twitter messages as part of the Social Media Mining for Health (SMM4H) 2021 Shared Task (Magge et al., 2021).

This work is based on the participation of our team in four subtasks of two tasks. Task 1 consists of three subtasks, namely 1a, 1b, and 1c each of which corresponds to classification, extraction, and normalization of ADEs. For Task 2, train, dev, and test sets include Russian tweets annotated with a binary label indicating the presence or absence of ADEs. For the 1b task, named entity recognition aims to detect the mentions of ADEs. Task 1c is designed as an end-to-end problem, intended to perform full evaluation of a system operating in real conditions: given a set of raw tweets, the system has to find the tweets that are mentioning ADEs, find the spans of the ADEs, and normalize them with respect to a given knowledge base (KB). These tasks are especially challenging due to specific characteristics of user-generated texts from social networks which are noisy, containing misspelled words, abbreviations, emojis, etc. The source code for our models is freely available[1].

The paper is organized as follows. We describe our experiments on the multilingual and multimodal classification of Russian and English tweets for the presence or absence of adverse effects in Section 2. In Section 3, we describe our pipeline for named entity recognition (NER) and medical concept normalization (MCN). Finally, we conclude this paper in Section 4.

2 Tasks 1a & 2: multilingual classification of tweets

The objective of Tasks 1a & 2 is to identify whether a tweet in English (Task 1a) or Russian (Task 2) mentions an adverse drug effect.

2.1 Data

For the English task, we used the original dev set provided by the organizers of the SMM4H 2021. For the Russian task, we sampled 1,000 non-repeating tweets from the original dev set as the new dev set and added the remaining tweets to the training set. Table 1 presents the statistics on Task 1a and Task 2 data. As can be seen from the table, the classes are highly imbalanced for both the English and the Russian corpora.

We preprocessed datasets for tasks 1a and 2 in a similar manner. During preprocessing, we: (i) replaced all URLs with the word "link"; (ii) replaced all user mentions with @username placeholder; (iii) replaced some emojis with a textual representation (e.g., laughing emojis with the word laughing; pill and syringe emojis with the corresponding

[1] https://github.com/Andoree/smm4h_2021_classification

Proceedings of the Sixth Social Media Mining for Health Workshop 2021, pages 39–43
June 10, 2021.

Dataset	# Tweets	# Positive samples (ADE presence)
Task 1a (English tweets)		
Train	17,385	1,235 (7.1%)
Dev	914	65 (7.1%)
Test	10,984	–
Task 2 (Russian tweets)		
Train	10,609	980 (9.2%)
Dev	1,000	92 (9.2%)
Test	9,095	–

Table 1: Task 1a and Task 2 data statistics

words); (iv) replaced ampersand's HTML representation "&" with "&". As training sets are highly imbalanced, we applied the positive class over-sampling so that each training batch contained roughly the same number of positive and negative samples. However, we did not observe a significant performance improvement for the Russian subtask, so we applied the technique for the English subtask only. Following (Miftahutdinov et al., 2020), for Task 2, we combined the English and the Russian training sets.

2.2 Models

For our experiments, we used neural models based on Bidirectional Encoder Representations from Transformers (BERT) (Devlin et al., 2019) as they have achieved state-of-the-art results in the biomedical domain. In particular, BERT-based models proved efficient at the SMM4H 2020 Shared Task (Gonzalez-Hernandez et al., 2020). We used the following BERT-based models:

(1) RoBERTa$_{large}$[2] (Liu et al., 2019), a modification of BERT that is pretrained on 160GB of English texts with dynamic masking;

(2) EnRuDR-BERT[3] (Tutubalina et al., 2020), pretrained on English (Tutubalina et al., 2017) and Russian (Tutubalina et al., 2020) corpora of 5M health-related texts in English and Russian;

(3) ChemBERTa[4] (Chithrananda et al., 2020), a RoBERTa$_{base}$-based model that is pretrained on compounds from ZINC (Irwin and Shoichet, 2005) and is designed for drug design, chemical modelling and molecular properties prediction.

[2]https://huggingface.co/roberta-large

[3]https://huggingface.co/cimm-kzn/enrudr-bert

[4]https://huggingface.co/seyonec/ChemBERTa_zinc250k_v2_40k

2.3 Experiments

For both tasks, we investigated the efficacy of the multimodal classification approach. For each tweet, we found its drug mentions, represented the chemical structure of each drug as a Simplified molecular-input line-entry system (SMILES) string, encoded the string using ChemBERTa, and took the final [CLS] embedding as drug embedding. Thus, we matched each tweet with a drug embedding. For tweets that contain no drug mentions, we encoded an empty string. We compared the following text-molecule combination strategies: (i) concatenation of the drug and the text embeddings, (ii) one cross-attention layer (Vaswani et al., 2017) from molecule encoder to text encoder. For concatenation architecture, we did not fine-tune ChemBERTa on the training set, whereas for cross-attention models, we trained both text and drug encoder.

For both Task 1a and Task 2, we adopted pretrained models from HuggingFace (Wolf et al., 2019) and fine-tuned them using PyTorch (Paszke et al., 2019). We trained each RoBERTa$_{large}$ model for 10 epochs with the learning rate of $1 * 10^{-5}$ using Adam optimizer (Kingma and Ba, 2014). We set batch size to 32 and maximum sequence size to 128. For EnRuDR-BERT we used the learning rate of $3 * 10^{-5}$, batch size to 64, and sequence to 128. For ChemBERTa, we used a sequence length of 256. For classification, we used a fully-connected network with one hidden layer, GeLU (Hendrycks and Gimpel, 2016) activation, a dropout probability of 0.3, and sigmoid as the final activation. To handle a high variance of BERT-based models' performance that varies across different initializations of classification layers, for each training setup, we trained 10 models and weighed their predictions. We tried two weighing strategies: (i) majority voting and (ii) sigmoid-based weighing. For (ii), we used predicted positive class probabilities to train a Scikit-learn's (Pedregosa et al., 2011) logistic regression on the validation set. For all experiments, we used a classification threshold of 0.5.

Table 2 shows the performance of our systems for Task 1a and Task 2 in terms of precision (P), recall (R), and F1-score (F1). Based on the results, we can draw the following conclusions. First, for the English task, the concatenation of text and chemical features increases the F1-score by 3%

Model set-up	P	R	F1
Task 1a (English tweets)			
RoBERTa	–	–	0.58
RoBERTa + ChemBERTa concatenation	–	–	**0.61**
* RoBERTa + ChemBERTa concatenation + over-sampling	0.55	0.68	**0.61**
* RoBERTa + ChemBERTa concatenation + over-sampling + sigmoid	0.59	0.56	0.58
Average scores provided by organizers	0.51	0.41	0.44
Task 2 (Russian tweets)			
EnRuDR-BERT + Ru train	–	–	0.55
EnRuDR-BERT + Ru train + ChemRoBERTa + cross-attention	–	–	0.54
EnRuDR-BERT + RuEn train	–	–	0.54
* EnRuDR-BERT + RuEn train + ChemRoBERTa cross-attention	0.58	0.57	**0.57**
* EnRuDR-BERT + RuEn train + ChemRoBERTa cross-attention + sigmoid	0.77	0.35	0.48
Average scores provided by organizers	0.55	0.56	0.51

Table 2: Text classification results on the SMM4H 2021 Task 1a and Task 2 test sets. For all set-ups except the ones with "sigmoid", we used majority voting. Our official submissions for the SMM4H 2021 are denoted by $*$.

compared to text-only classification. Second, for the Russian task, neither the bilingual approach nor the use of chemical features shows a performance improvement when used separately, but the joint use of bilingual data and cross-modality with cross-attention results in an F1-score growth of 2% compared to text-only monolingual classification. Third, the results of this year showed a smaller gap between F1-scores on Russian and English test sets than last year.

3 Tasks 1b & 1c: extraction and normalization of ADEs

The 1b task's objective is to detect ADE mentions. Task 1c is designed as an end-to-end task. Systems have a free-form tweet as input and should be able to produce a set of extracted medical concepts. For this task, we develop a pipeline that (i) first detect ADE mentions and then (ii) link extracted ADEs to the concepts from the medical dictionary for regulatory activities (MedDRA) (Brown et al., 1999).

Following the best results in SMM4H 2020 Task 3 (Miftahutdinov et al., 2020), we utilize a EnDR-BERT model[5] with dictionary based features for the named entity recognition (NER) task. We adopted the dictionaries from (Miftahutdinov et al., 2017). As in the best solution of the SMM4H 2020 Task 3, we adopted extra training data for the NER task, we used the CSIRO Adverse Drug Event Corpus (CADEC) (Karimi et al., 2015) and COMETA corpus (Basaldella et al., 2020).

For the normalization task, we applied two models: (i) a classifier (Miftahutdinov et al., 2020; Miftahutdinov and Tutubalina, 2019), (ii) a novel neural model based on similarity distance of BERT vectors of concepts (Miftahutdinov et al., 2021). Following (Miftahutdinov et al., 2020), we utilize additional data for training. Other corpora are filtered to match a vocabulary of the SMM4H 2021 train set. We combined two models based on a threshold. For instance, given (i) prediction c_{bs} from from BERT-based similarity method with the distance equals to d and (ii) prediction c_{clf} from the classification approach, the final prediction is set to c_{bs}, if d is less than a threshold, and to c_{clf}, otherwise. For more detailed description of NER and end-to-end entity linking model please refer to (Miftahutdinov et al., 2020).

Table 3 shows a comparison of the model to the official average scores computed using the participants' submissions. Our NER model achieved below average results (40% vs 42%). We believe that the results are related to additional training of the model on non-target texts (reviews). Yet, with lower results in Task 1b and the top ranked results in Task 1c, it becomes clear that that the advantage of our pipeline is the two-component model for medical concept normalization. To sum up, the pipeline ranked first at SMM4H 2021 Task 1c and obtained the F1 score of 29% on extraction of MedDRA concepts.

[5] https://huggingface.co/cimm-kzn/endr-bert

Run name	P	R	F1
ADE Detection Evaluation (Task 1b)			
KFU NLP Team	0.42	0.38	0.40
Average scores	0.49	0.46	**0.42**
End-to-End Evaluation (Task 1c)			
KFU NLP Team	0.30	0.28	**0.29**
Average scores	0.23	0.22	0.22

Table 3: Performance of our models in SMM4H 2021 Task 1b and 1c (official results).

4 Conclusion

In this work, we have explored an application of domain-specific BERT models pretrained on health-related user reviews in English and Russian to the task of multilingual and multimodal text classification, extraction, and normalization of adverse drug effects. Our experiments show that multimodal architecture for classification of tweets outperforms other strong baselines and text classifiers. Besides, our BERT-based pipeline for extraction on MedDRA concepts ranked 1st in Task 1c.

We foresee two directions for future work. First, future research will explore how different drug representation models and pretraining approaches affect classification performance. Second, a potential direction is to verify the efficacy of multimodal classification for languages other than Russian and English.

Acknowledgements

The work on neural models has been supported by the Russian Science Foundation grant # 18-11-00284. Part of NER experiments was carried out by Elena Tutubalina on a lab computer (workstation) at HSE University within the framework of the HSE University Basic Research Program.

References

Marco Basaldella, Fangyu Liu, Ehsan Shareghi, and Nigel Collier. 2020. COMETA: A corpus for medical entity linking in the social media. In *Proceedings of the 2020 Conference on Empirical Methods in Natural Language Processing (EMNLP)*, pages 3122–3137, Online. Association for Computational Linguistics.

Elliot G Brown, Louise Wood, and Sue Wood. 1999. The medical dictionary for regulatory activities (meddra). *Drug safety*, 20(2):109–117.

Seyone Chithrananda, Gabe Grand, and Bharath Ramsundar. 2020. Chemberta: Large-scale self-supervised pretraining for molecular property prediction. *arXiv preprint arXiv:2010.09885*.

Jacob Devlin, Ming-Wei Chang, Kenton Lee, and Kristina Toutanova. 2019. BERT: Pre-training of deep bidirectional transformers for language understanding. In *Proceedings of the 2019 Conference of the North American Chapter of the Association for Computational Linguistics: Human Language Technologies, Volume 1 (Long and Short Papers)*, pages 4171–4186, Minneapolis, Minnesota. Association for Computational Linguistics.

Graciela Gonzalez-Hernandez, Ari Z. Klein, Ivan Flores, Davy Weissenbacher, Arjun Magge, Karen O'Connor, Abeed Sarker, Anne-Lyse Minard, Elena Tutubalina, Zulfat Miftahutdinov, and Ilseyar Alimova, editors. 2020. *Proceedings of the Fifth Social Media Mining for Health Applications Workshop & Shared Task*. Association for Computational Linguistics, Barcelona, Spain (Online).

Dan Hendrycks and Kevin Gimpel. 2016. Gaussian error linear units (gelus). *arXiv preprint arXiv:1606.08415*.

John J Irwin and Brian K Shoichet. 2005. Zinc- a free database of commercially available compounds for virtual screening. *Journal of chemical information and modeling*, 45(1):177–182.

Sarvnaz Karimi, Alejandro Metke-Jimenez, Madonna Kemp, and Chen Wang. 2015. Cadec: A corpus of adverse drug event annotations. *Journal of biomedical informatics*, 55:73–81.

Diederik P. Kingma and Jimmy Ba. 2014. Adam: A method for stochastic optimization. *CoRR*, abs/1412.6980.

Yinhan Liu, Myle Ott, Naman Goyal, Jingfei Du, Mandar Joshi, Danqi Chen, Omer Levy, Mike Lewis, Luke Zettlemoyer, and Veselin Stoyanov. 2019. Roberta: A robustly optimized bert pretraining approach. *arXiv preprint arXiv:1907.11692*.

Arjun Magge, Ari Klein, Ivan Flores, Ilseyar Alimova, Mohammed Ali Al-garadi, Antonio Miranda-Escalada, Zulfat Miftahutdinov, Eulàlia Farré-Maduell, Salvador Lima López, Juan M Banda, Karen O'Connor, Abeed Sarker, Elena Tutubalina, Martin Krallinger, Davy Weissenbacher, and Graciela Gonzalez-Hernandez. 2021. Overview of the sixth social media mining for health applications (# smm4h) shared tasks at naacl 2021. In *Proceedings of the Sixth Social Media Mining for Health Applications Workshop & Shared Task*.

Z.Sh. Miftahutdinov, E.V. Tutubalina, and A.E. Tropsha. 2017. Identifying disease-related expressions in reviews using conditional random fields. *Komp'juternaja Lingvistika i Intellektual'nye Tehnologii*, 1(16):155–166.

Zulfat Miftahutdinov, Artur Kadurin, Roman Kudrin, and Elena Tutubalina. 2021. Drug and disease interpretation learning with biomedical entity representation transformer. *Proceedings of the 43rd European Conference on Information Retrieval.*

Zulfat Miftahutdinov, Andrey Sakhovskiy, and Elena Tutubalina. 2020. KFU NLP team at SMM4H 2020 tasks: Cross-lingual transfer learning with pretrained language models for drug reactions. In *Proceedings of the Fifth Social Media Mining for Health Applications Workshop & Shared Task*, pages 51–56, Barcelona, Spain (Online). Association for Computational Linguistics.

Zulfat Miftahutdinov and Elena Tutubalina. 2019. Deep neural models for medical concept normalization in user-generated texts. In *Proceedings of the 57th Annual Meeting of the Association for Computational Linguistics: Student Research Workshop*, pages 393–399.

Adam Paszke, Sam Gross, Francisco Massa, Adam Lerer, James Bradbury, Gregory Chanan, Trevor Killeen, Zeming Lin, Natalia Gimelshein, Luca Antiga, Alban Desmaison, Andreas Kopf, Edward Yang, Zachary DeVito, Martin Raison, Alykhan Tejani, Sasank Chilamkurthy, Benoit Steiner, Lu Fang, Junjie Bai, and Soumith Chintala. 2019. Pytorch: An imperative style, high-performance deep learning library. In H. Wallach, H. Larochelle, A. Beygelzimer, F. d'Alché-Buc, E. Fox, and R. Garnett, editors, *Advances in Neural Information Processing Systems 32*, pages 8024–8035. Curran Associates, Inc.

Fabian Pedregosa, Gaël Varoquaux, Alexandre Gramfort, Vincent Michel, Bertrand Thirion, Olivier Grisel, Mathieu Blondel, Peter Prettenhofer, Ron Weiss, Vincent Dubourg, et al. 2011. Scikit-learn: Machine learning in python. *the Journal of machine Learning research*, 12:2825–2830.

Elena Tutubalina, Ilseyar Alimova, Zulfat Miftahutdinov, Andrey Sakhovskiy, Valentin Malykh, and Sergey Nikolenko. 2020. The russian drug reaction corpus and neural models for drug reactions and effectiveness detection in user reviews. *Bioinformatics*. Btaa675.

EV Tutubalina, Z Sh Miftahutdinov, RI Nugmanov, TI Madzhidov, SI Nikolenko, IS Alimova, and AE Tropsha. 2017. Using semantic analysis of texts for the identification of drugs with similar therapeutic effects. *Russian Chemical Bulletin*, 66(11):2180–2189.

Ashish Vaswani, Noam Shazeer, Niki Parmar, Jakob Uszkoreit, Llion Jones, Aidan N Gomez, Łukasz Kaiser, and Illia Polosukhin. 2017. Attention is all you need. In *Proceedings of the 31st International Conference on Neural Information Processing Systems*, pages 6000–6010.

Thomas Wolf, Lysandre Debut, Victor Sanh, Julien Chaumond, Clement Delangue, Anthony Moi, Pierric Cistac, Tim Rault, Rémi Louf, Morgan Funtowicz, et al. 2019. Huggingface's transformers: State-of-the-art natural language processing. *arXiv preprint arXiv:1910.03771*.

Transformer-based Multi-Task Learning for Adverse Effect Mention Analysis in Tweets

George-Andrei Dima[1,2], Dumitru-Clementin Cercel[1], Mihai Dascalu[1]
University Politehnica of Bucharest, Faculty of Automatic Control and Computers[1]
Military Technical Academy Ferdinand I[2]
andrei.dima@mta.ro, {dumitru.cercel, mihai.dascalu}@upb.ro

Abstract

While social media gains a broader traction, valuable insights and opinions on various topics representative for a wider audience can be automatically extracted using state-of-the-art Natural Language Processing techniques. Of particular interest in the healthcare domain are adverse drug effects, which may be introduced in online posts, and can be effectively centralized and investigated. This paper presents our Multi-Task Learning architecture using pretrained Transformer-based language models employed for the Social Media Mining for Health Applications Shared Task 2021, where we tackle the three subtasks of Task 1, namely: classification of tweets containing adverse effects (subtask 1a), extraction of text spans containing adverse effects (subtask 1b), and adverse effects resolution (subtask 1c). Our best performing model ranked first on the test set at subtask 1b with an F_1-*score* of 51% (P = 51%; R = 51%). Promising results were obtained on subtask 1a (F_1-*score* = 44%; P = 45%; R = 43%), whereas subtask 1c was by far the most difficult task and an F_1-*score* of only 17% (P = 17%; R = 18%) was obtained.

1 Introduction

Information extraction from social media is widely studied nowadays, as platforms like Facebook, Twitter, Instagram, or Reddit become the main place for people to share their opinions and experiences. Concurrently, a wide range of applications with completely different topics arises with the current advances in Natural Language Processing (NLP), as the volume of posted information has become impossible to be manually analysed. For example, the Social Media Mining for Health (SMM4H) Applications Shared Task (Sarker and Gonzalez-Hernandez, 2017) is focused on health applications and introduces a dataset of annotated tweets with the aim to analyse adverse drug effects (ADE) mentioned by users. This year's edition (Magge et al., 2021) proposed eight different tasks, out of which we focused on the first task entitled *Classification, Extraction and Normalization of Adverse Effect mentions in English tweets*. This task was further divided into three subtasks, as follows. *Subtask 1a* was a binary classification task of tweets, focused on identifying whether the message contains ADE or not. *Subtask 1b* was a named entity recognition task on top of subtask 1a, centered on extracting the span of text containing the ADE of the medication. *Subtask 1c* was a named entity resolution task on top of both previous subtasks, aimed at predicting the normalized concept of the extracted adverse effect from the preferred terms included in the Medical Dictionary for Regulatory Activities (MedDRA)[1].

All three subtasks are addressed simultaneously using a Multi-Task Learning (MTL) architecture (Caruana, 1997) that leverages acquired knowledge from one subtask to another. Furthermore, we approached the challenge of unbalanced classes in the first subtask by considering class weights and by augmenting the training data set.

The paper is structured as follows. The second section describes previous work that inspired our solution, while the third section presents our employed method. The fourth section presents the results of our work, followed by discussions and the final section that summarizes our findings and presents future research paths.

2 Related Work

2.1 Health-Related Applications

Given that Task 1 was present in previous editions of the SMM4h shared task (Weissenbacher et al., 2018, 2019; Klein et al., 2020), several approaches were employed to address its challenges. For example, the winning team from 2019 (Miftahutdinov

[1]https://www.meddra.org/

Proceedings of the Sixth Social Media Mining for Health Workshop 2021, pages 44–51
June 10, 2021.

et al., 2019) used an ensemble of BioBERT-CRF models for the ADE extraction task, while addressing the resolution task as a classification. The system proposed by Miftahutdinov et al. (2020) ranked first at the end-to-end 2020 competition using the pretrained EnDR-BERT (Tutubalina et al., 2020) and the CSIRO Adverse Drug Event Corpus (CADEC) (Karimi et al., 2015) for further training the model. In addition, Dima et al. (2020) showed that bidirectional Transformers trained using class weighting, together with ensembles that combine various configurations, achieve an F1-score of .705 on the dataset made available for that edition of the competition.

2.2 MTL-Based Methods

Multi-Task Learning represents a training strategy where a shared model is simultaneously learning multiple tasks. Ruder (2017) analysed the techniques applied in MTL and compared the *hard parameter sharing* and *soft parameter sharing* paradigms, concluding that the former is still pervasive in nowadays approaches. MTL proved to fasten the convergence and to improve the model performance in a variety of NLP applications, including named entity recognition (Aguilar et al., 2018), fake news detection (Wu et al., 2019), multilingual offensive language identification (Chen et al., 2020b), sentiment analysis (Zaharia et al., 2020), humor classification (Vlad et al., 2020), recommender systems (Tang et al., 2020), and even question answering (Kongyoung et al., 2020). MTL also increases performance in conjunction with semi-supervised learning (Liu et al., 2007), curriculum learning (Dong et al., 2017), sequence-to-sequence (Zaremoodi and Haffari, 2018), reinforcement learning (Gupta et al., 2020), and adversarial learning (Liu et al., 2017).

3 Method

3.1 Corpus

The SMM4H 2021 Task 1 dataset included 17,385 training samples out of which 1,235 (7.10%) belong to the positive class (i.e., contain ADE), as well as 915 samples in the development set out of which 65 are labeled as positive; hence, a challenge consists of the unbalanced distribution of the two classes.

Subtask 1c required labeling the extracted text span with the corresponding MedDRA term; the number of possible labels exceeds 23,000. Only 476 labels are present in the training set, denoting that most labels are not covered at all. Additionally, the number of appearances of each ID has a long-tail distribution (see Figure 1), with some IDs being present in more than 60 examples and most IDs occurring in less than 4 examples.

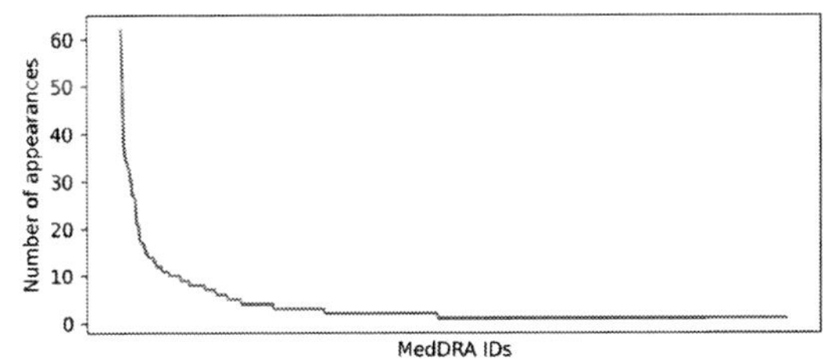

Figure 1: Histogram of MedDRA IDs present in the training set.

3.2 Multi-Task Learning Neural Architecture

A MTL architecture based on hard parameter sharing (Ruder, 2017) was employed for Task 1 (see Figure 2). Given that all three subtasks are highly related, our assumption was that knowledge acquired while learning one subtask would help in increasing performance on the other two. Three modules are added on top of BioBERT (Lee et al., 2020): (a) the *Classifier* - a binary classifier for tweet classification in subtask 1a, b) the *Extractor* - a named entity recognition layer for ADE span extraction in subtask 1b, and c) the *Normalizer* - a multi-class classifier for span resolution in subtask 1c. All three modules share the same pre-trained BERT encoder; the first 11 layers out of 12 were frozen, whereas the last layer was kept as a shared trainable encoder.

The training dataset was processed in the following manner. The positive tweets from the training set were selected for subtask 1b, and each token was tagged with either "O" (outside adverse effect) or "AE" (adverse effect entity). Two approaches were considered for subtask 1c: (a) create a dataset using the spans labeled with their corresponding PTID, and (b) concatenate the span tokens with the corresponding tweets as: *[CLS] <ADE span > [SEP] <entire tweet> [SEP]*. The second approach aimed to leverage context information in the MedDRA ID prediction.

The modules for the three subtasks were trained in parallel. At each training step, a batch from the training datasets was randomly chosen with the

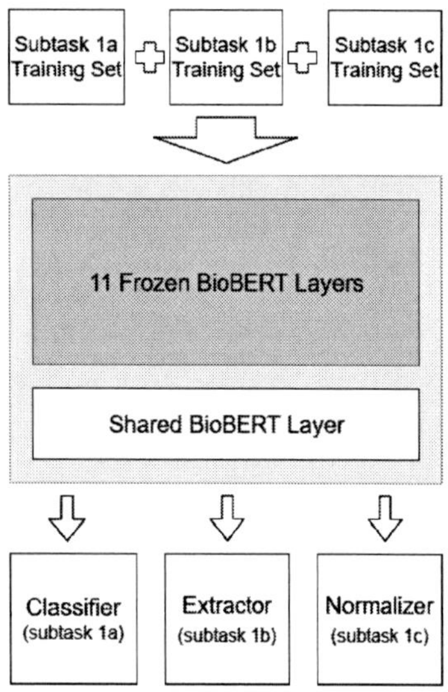

Figure 2: The overall architecture of the proposed multi-task framework.

following probability:

$$p_i = \frac{\frac{size(D_i)}{size(b_i)}}{(\sum_{k=1}^{n} \frac{size(D_k)}{size(b_k)})} \quad (1)$$

where D_i represents the dataset for subtask i, and b_i represents a mini-batch from D_i.

All three subtasks minimize cross-entropy loss (see Equation 2), whereas only subtask 1a considers values different from one for weights w_j:

$$L_{Task} = -w_j \sum_i (y_i \log \hat{y}_i + (1 - y_i) \log(1 - \hat{y}_i)) \quad (2)$$

The final loss minimized at each step k of Algorithm 1 is expressed in Equation 3:

$$L_k = t_k^1 L_{cls} + t_k^2 L_{ner} + t_k^3 L_{norm}, \quad (3)$$

where $t_k^i = 1$ when task i is in training, or $t_k^i = 0$ otherwise.

Algorithm 1: Multi-task training algorithm

Initialize model parameters Θ from BioBERT using transfer learning;
Compute probabilities p_i for each subtask i using Equation 1;
Shuffle and pack datasets into mini-batches D_1, D_2, D_3;
for N *training steps* **do**
    Choose task i with probability p_i;
    if D_i *is empty* **then**
        Shuffle and re-pack D_i;
    end
    Randomly choose mini-batch b from D_i;
    Compute task specific loss L_i on b;
    Update Θ using L_i;
end

Algorithm 2 describes the processing pipeline which begins by passing the input tweet through the *Classifier*. If a tweet is labeled as not containing an adverse effect, the label is memorized and the flow stops. Otherwise, the tweet is passed to the *Extractor* and, afterwards, to the *Normalizer*. Line 10 highlights that the input tweet can also be used besides the text span that contains the ADE, in order to leverage the context information in predicting the MedDRA ID; this feature is optional and can be deactivated in certain configurations. Moreover, the feedback loop from the *Extractor* considering the label of the tweet (line 12) is optional.

Algorithm 2: Task 1 prediction algorithm

1 Initialize $Classifier$;
2 Initialize $Extractor$;
3 Initialize $Normalizer$;
4 Load dataset D;
5 **for** $tweet$ *in* D **do**
6     $label \leftarrow Classifier(tweet)$;
7   **if** $label$ *is* ADE **then**
8         $S \leftarrow Extractor(tweet)$;
9   **if** S *is* ***not*** *empty* **then**
10            $I \leftarrow Normalizer(span, tweet)$ for each $span$ in S;
11   **else**
12   Change $label$ to $NoADE$;
13   **end**
14   **end**
15   Save $(label, S, I)$;
16 **end**

3.3 Implementation Details

Language Models: We experimented with BERT-base (Devlin et al., 2019) and with the domain-specific Transformers, namely BioBERT and BioClinicalBERT (Alsentzer et al., 2019). After a preliminary fine-tuning on the subtask 1a, the most promising results were obtained by BioBERT.

Given the limited resources available, we kept it as the default pre-trained solution in all further experiments.

Hyperparameters: All three modules (*Classifier*, *Extractor*, and *Normalizer*) were trained with a learning rate of $5e-5$. Batch sizes of 64 were used for subtask 1a that had most entries, while batch sizes of 16 were considered for subtasks 1b, and 1c in which only positive samples are considered. Training was performed for 30 epochs, computing the performance on the validation set after each epoch and saving the system that performed best.

Class Weights: The class unbalance problem from subtask 1a is addressed using the weighted version of the cross-entropy loss. The weights of the two classes were computed using the *balanced* heuristic (King and Zeng, 2001) from the scikit-learn library (Pedregosa et al., 2011).

Augmented Training Dataset: Another explored solution for the unbalance in subtask 1a consists in augmenting the poorly represented class (the positive class). We leverage the predefined augmentation approaches integrated into the TextAttack library (Morris et al., 2020). New positive examples are generated by char swapping, by replacing words with synonyms from the WordNet thesaurus (Miller, 1995), and by using methods from the CheckList testing - i.e., transformations like location replacement or number alteration (Ribeiro et al., 2020). Five positive examples are automatically added for each initial positive sample, thus increasing the proportion of the poorly represented class from 7% to almost 45%.

Class Number Reduction for the Normalizer: We considered subtask 1c a multi-class classification task where the *Normalizer* module receives as input the text span containing an ADE (i.e., the output of the *Extractor* module) and classifies the span into one of the classes (i.e., MedDRA PTIDs) present in the training set. The distribution of the 476 MedDRA IDs influenced us to reduce the number of classes. As such, the final classifier considers only the most frequent 108 PTIDs (i.e., IDs that appear more than three times in the training dataset). There were too few examples to properly generalize for all PTIDs; however, the module covers only 69.5% of the training samples.

4 Results

Four configurations were compared in terms of performance. The first configuration (*MTL*) is a baseline relying on the previously described MTL architecture. Weighted binary cross-entropy loss and feedback from the *Extractor* to the *Normalizer* (Line 12 from Algorithm 2) are enabled, but the *Normalizer* uses only the ADE span, without the entire tweet (Line 10 from Algorithm 2).

The second configuration (*MTL + BoostingEnsemble*) starts from *MTL*, but instead of the simple *Classifier* model, it uses an ensemble of three models trained in a boosting manner. The first classifier (*Classifier1*) is identical to the classifier from the first configuration. The second classifier (*Classifier2*) was trained on a modified training set in which the miss-classifies examples from *Classifier1* are over-sampled by a factor of three, whereas the correctly classified examples are down-sampled by the same factor. The third classifier, *Classifier3*, is also trained on a modified training set in which examples with different results from *Classifier1* and *Classifier2* are over-sampled by a factor of three, while the rest of the examples are down-sampled by the same factor.

The third configuration, denoted *MTL + EnhancedEnsemble*, further tries to improve the performance of the second configuration by adding two more classifiers to the ensemble, *Classifier4* and *Classifier5*, trained now on the augmented training set while considering equivalent over- and down-sampling approaches.

The fourth configuration, namely *MTL + EnhancedNormalizer*, is similar to the first configuration, but with the *Normalizer* is trained on both the ADE span and the entire tweet.

Table 1 introduces the comparative results for all configurations. While considering the development dataset, *MTL + EnhancedEnsemble* obtains the best performance for subtasks 1a and 1b, while *MTL + EnhancedNormalizer* has the highest *F1-score* for subtask 1c. In terms of the test dataset, *MTL* has the highest F1-score for subtask 1a - although the other configurations gain a boost in precision, recall is negatively influenced; *MTL + BoostingEnsemble* has the best performance on subtask 1b, whereas *MTL + EnhancedNormalizer* remains the best configuration for subtask 1c. Although *MTL + EnhancedEnsemble* has better results while integrating the augmented dataset, there are no improvements on the test dataset.

Model	Subtask	Development			Test		
		P (%)	R (%)	F_1 (%)	P (%)	R (%)	F_1 (%)
MTL	1a	58.9	66.1	62.3	45.2	43.4	**44.3**
	1b	44.3	64.1	52.4	44.2	53.6	48.5
	1c	14.2	21.8	17.2	14.5	18.7	16.4
MTL + BoostingEnsemble*	1a	61.7	64.6	63.1	49.1	39.3	43.7
	1b	44.1	61.9	51.5	51.4	51.4	**51.4**
	1c	15.5	22.9	18.5	16.0	17.0	16.0
MTL + EnhancedEnsemble*	1a	65.1	62.1	**63.5**	48.8	36.6	42.0
	1b	50.8	62.3	**56.0**	51.4	49.1	50.0
	1c	15.7	20.6	17.9	15.8	16.0	16.0
MTL + EnhancedNormalizer	1c	17.2	24.1	**20.0**	16.9	17.9	**17.4**

Table 1: Evaluation of configurations for each subtask of SMM4H Task 1.
* marks the official submissions.

5 Discussions

Table 2 introduces classification problems that provide additional insights on how our *MTL + BoostingEnsemble* model works. Overall, it correctly extracts and classifies most text spans containing usual words for adverse effects (e.g., "sick") but, it has occasional difficulties in distinguishing between the desired effect of a medication and its adverse effects. For instance, in the first example from Table 2, our model does not make the association that the described medication is supposed to help the subject sleep, but, in contrast, it assumes sleepiness as an adverse effect.

Another limitation of our method is highlighted in the second example. The MedDRA term of *Slurred speech* is a rather rare label, not even present in the training set. Even though our system correctly extracts the span containing the adverse effect, it is unable to correctly predict the *ptid*.

The false positive example of "drunk" labeled as *Drunk like effect* shows that our model finds it hard to discern appearances from facts. A similar bias can be observed in the third example, where the model fails to extract the spans "sleep" and "stomach is a cement mixer" most likely because it learned that interrogations ask about adverse effects rather than offer information about them.

The fourth example denotes subtle errors, like grasping the difference between the MedDRA terms of *Sleepiness* and *Somnolence*, which are likely to be mislabeled even by humans.

While considering the differences between development and test set performances, another limi-

Annotated sample	MedDRA	Model prediction	MedDRA
...trazodone, it takes the light right outta your eyes...		...trazodone, it takes the light right outta your eyes ...	*Sleepiness*
one of the things i hate most about quetiapine is when i take it for the first few hours i slur my words, so people assume i'm merely drunk.	*Slurred speech*	one of the things i hate most about quetiapine is when i take it for the first few hours i slur my words, so people assume i'm merely drunk .	*Fluid retention, Drunk-like effect*
ciprofloxacin: how do you expect to sleep when your stomach is a cement mixer ?	*Sleeplessness, Stomach perforation*	ciprofloxacin: how do you expect to sleep when your stomach is a cement mixer?	
just woke up. since i started on the higher dose of quetiapine i'm sleeping even more ...; i feel knackered when i wake .	*Sleepiness, Groggy on awakening*	just woke up. since i started on the higher dose of quetiapine i'm sleeping even more ...; i feel knackered when i wake.	*Somnolence, Feeling stoned*

Table 2: Examples from the validation set obtained using *MTL + BoostingEnsemble*. Note that the MedDRA IDs were replaced by their *Preferred Terms*.

tation emerges, namely that our configurations did not generalized as expected on the test set for subtask 1a. This is argued by the reduced development set which contains only 5% of the provided labeled examples, coupled with our training procedure of always saving the model at its best validation score.

6 Conclusions and Future Work

We introduced a Transformer-based Multi-Task Learning architecture employed for Task 1 from the Social Media Mining for Health Applications Shared Task 2021. Task 1 was concerned with the classification of tweets incorporating adverse effects of medication and, for the positive tweets, with the extraction and normalization of the adverse effects. We started from a pretrained domain-specific BERT language model (i.e., BioBERT) which was further finetuned in a multi-task setting. A hard parameter sharing MTL model was trained on the three subtasks of SMM4H Task 1. Furthermore, class weights and data augmentation were considered to overcome the problem of the unbalanced dataset from subtask 1a.

Our model achieved the highest score for subtask 1b (i.e., adverse effect span detection) with an F_1-score of 51%, arguing that MTL can enhance adverse effect extraction from social media posts. In terms of future work, adversarial training (Miyato et al., 2018; Chen et al., 2020a) will be considered to improve the robustness of our approach.

References

Gustavo Aguilar, Adrian Pastor López-Monroy, Fabio A González, and Thamar Solorio. 2018. Modeling noisiness to recognize named entities using multitask neural networks on social media. In *Proceedings of the 2018 Conference of the North American Chapter of the Association for Computational Linguistics: Human Language Technologies, Volume 1 (Long Papers)*, pages 1401–1412.

Emily Alsentzer, John Murphy, William Boag, Wei-Hung Weng, Di Jindi, Tristan Naumann, and Matthew McDermott. 2019. Publicly available clinical bert embeddings. In *Proceedings of the 2nd Clinical Natural Language Processing Workshop*, pages 72–78.

Rich Caruana. 1997. Multitask learning. *Machine learning*, 28(1):41–75.

Kejiang Chen, Yuefeng Chen, Hang Zhou, Xiaofeng Mao, Yuhong Li, Yuan He, Hui Xue, Weiming Zhang, and Nenghai Yu. 2020a. Self-supervised adversarial training. In *ICASSP 2020-2020 IEEE International Conference on Acoustics, Speech and Signal Processing (ICASSP)*, pages 2218–2222. IEEE.

Po Chun Chen, Hen-Hsen Huang, and Hsin-Hsi Chen. 2020b. Ntu_nlp at semeval-2020 task 12: Identifying offensive tweets using hierarchical multi-task learning approach. In *Proceedings of the Fourteenth Workshop on Semantic Evaluation*, pages 2105–2110.

Jacob Devlin, Ming-Wei Chang, Kenton Lee, and Kristina Toutanova. 2019. Bert: Pre-training of deep bidirectional transformers for language understanding. In *Proceedings of the 2019 Conference of the North American Chapter of the Association for Computational Linguistics: Human Language Technologies, Volume 1 (Long and Short Papers)*, pages 4171–4186.

George-Andrei Dima, Andrei-Marius Avram, and Dumitru-Clementin Cercel. 2020. Approaching smm4h 2020 with ensembles of bert flavours. In *Proceedings of the Fifth Social Media Mining for Health Applications Workshop & Shared Task*, pages 153–157.

Qi Dong, Shaogang Gong, and Xiatian Zhu. 2017. Multi-task curriculum transfer deep learning of clothing attributes. In *2017 IEEE Winter Conference on Applications of Computer Vision (WACV)*, pages 520–529. IEEE.

Deepak Gupta, Hardik Chauhan, Ravi Tej Akella, Asif Ekbal, and Pushpak Bhattacharyya. 2020. Reinforced multi-task approach for multi-hop question generation. In *Proceedings of the 28th International Conference on Computational Linguistics*, pages 2760–2775.

Sarvnaz Karimi, Alejandro Metke-Jimenez, Madonna Kemp, and Chen Wang. 2015. Cadec: A corpus of adverse drug event annotations. *Journal of biomedical informatics*, 55:73–81.

Gary King and Langche Zeng. 2001. Logistic regression in rare events data. *Political analysis*, 9(2):137–163.

Ari Z. Klein, Ilseyar Alimova, Ivan Flores, Arjun Magge, Zulfat Miftahutdinov, Anne-Lyse Minard, Karen O'Connor, Abeed Sarker, Elena Tutubalina, Davy Weissenbacher, and Graciela Gonzalez-Hernandez. 2020. Overview of the fifth social media mining for health applications (#smm4h) shared tasks at coling 2020. In *Proceedings of the Fifth Social Media Mining for Health Applications (#SMM4H) Workshop & Shared Task*.

Sarawoot Kongyoung, Craig Macdonald, and Iadh Ounis. 2020. Multi-task learning using dynamic task weighting for conversational question answering. In *Proceedings of the 5th International Workshop on Search-Oriented Conversational AI (SCAI)*, pages 17–26.

Jinhyuk Lee, Wonjin Yoon, Sungdong Kim, Donghyeon Kim, Sunkyu Kim, Chan Ho So, and Jaewoo Kang. 2020. Biobert: a pre-trained biomedical language representation model for biomedical text mining. *Bioinformatics*, 36(4):1234–1240.

Pengfei Liu, Xipeng Qiu, and Xuan-Jing Huang. 2017. Adversarial multi-task learning for text classification. In *Proceedings of the 55th Annual Meeting of the Association for Computational Linguistics (Volume 1: Long Papers)*, pages 1–10.

Qiuhua Liu, Xuejun Liao, and Lawrence Carin. 2007. Semi-supervised multitask learning. In *Proceedings of the 20th International Conference on Neural Information Processing Systems*, pages 937–944.

Arjun Magge, Ari Klein, Ivan Flores, Ilseyar Alimova, Mohammed Ali Al-garadi, Antonio Miranda-Escalada, Zulfat Miftahutdinov, Eulàlia Farré-Maduell, Salvador Lima López, Juan M Banda, Karen O'Connor, Abeed Sarker, Elena Tutubalina, Martin Krallinger, Davy Weissenbacher, and Graciela Gonzalez-Hernandez. 2021. Overview of the sixth social media mining for health applications (# smm4h) shared tasks at naacl 2021. In *Proceedings of the Sixth Social Media Mining for Health Applications Workshop & Shared Task*.

Zulfat Miftahutdinov, Ilseyar Alimova, and Elena Tutubalina. 2019. Kfu nlp team at smm4h 2019 tasks: Want to extract adverse drugs reactions from tweets? bert to the rescue. In *Proceedings of the fourth social media mining for health applications (# SMM4H) workshop & shared task*, pages 52–57.

Zulfat Miftahutdinov, Andrey Sakhovskiy, and Elena Tutubalina. 2020. Kfu nlp team at smm4h 2020 tasks: Cross-lingual transfer learning with pretrained language models for drug reactions. In *Proceedings of the Fifth Social Media Mining for Health Applications Workshop & Shared Task*, pages 51–56.

George A Miller. 1995. Wordnet: a lexical database for english. *Communications of the ACM*, 38(11):39–41.

Takeru Miyato, Shin-ichi Maeda, Masanori Koyama, and Shin Ishii. 2018. Virtual adversarial training: a regularization method for supervised and semi-supervised learning. *IEEE transactions on pattern analysis and machine intelligence*, 41(8):1979–1993.

John Morris, Eli Lifland, Jin Yong Yoo, Jake Grigsby, Di Jin, and Yanjun Qi. 2020. Textattack: A framework for adversarial attacks, data augmentation, and adversarial training in nlp. In *Proceedings of the 2020 Conference on Empirical Methods in Natural Language Processing: System Demonstrations*, pages 119–126.

F. Pedregosa, G. Varoquaux, A. Gramfort, V. Michel, B. Thirion, O. Grisel, M. Blondel, P. Prettenhofer, R. Weiss, V. Dubourg, J. Vanderplas, A. Passos, D. Cournapeau, M. Brucher, M. Perrot, and E. Duchesnay. 2011. Scikit-learn: Machine learning in Python. *Journal of Machine Learning Research*, 12:2825–2830.

Marco Tulio Ribeiro, Tongshuang Wu, Carlos Guestrin, and Sameer Singh. 2020. Beyond accuracy: Behavioral testing of nlp models with checklist. In *Proceedings of the 58th Annual Meeting of the Association for Computational Linguistics*, pages 4902–4912.

Sebastian Ruder. 2017. An overview of multi-task learning in deep neural networks. *arXiv preprint arXiv:1706.05098*.

Abeed Sarker and Graciela Gonzalez-Hernandez. 2017. Overview of the second social media mining for health (smm4h) shared tasks at amia 2017. *Training*, 1(10,822):1239.

Hongyan Tang, Junning Liu, Ming Zhao, and Xudong Gong. 2020. Progressive layered extraction (ple): A novel multi-task learning (mtl) model for personalized recommendations. In *Fourteenth ACM Conference on Recommender Systems*, pages 269–278.

Elena Tutubalina, Ilseyar Alimova, Zulfat Miftahutdinov, Andrey Sakhovskiy, Valentin Malykh, and Sergey Nikolenko. 2020. The russian drug reaction corpus and neural models for drug reactions and effectiveness detection in user reviews. *Bioinformatics*.

George-Alexandru Vlad, George-Eduard Zaharia, Dumitru-Clementin Cercel, Costin Chiru, and Stefan Trausan-Matu. 2020. Upb at semeval-2020 task 8: Joint textual and visual modeling in a multi-task learning architecture for memotion analysis. In *Proceedings of the Fourteenth Workshop on Semantic Evaluation*, pages 1208–1214.

Davy Weissenbacher, Abeed Sarker, Arjun Magge, Ashlynn Daughton, Karen O'Connor, Michael Paul, and Graciela Gonzalez. 2019. Overview of the fourth social media mining for health (smm4h) shared tasks at acl 2019. In *Proceedings of the Fourth Social Media Mining for Health Applications (# SMM4H) Workshop & Shared Task*, pages 21–30.

Davy Weissenbacher, Abeed Sarker, Michael Paul, and Graciela Gonzalez. 2018. Overview of the third social media mining for health (smm4h) shared tasks at emnlp 2018. In *Proceedings of the 2018 EMNLP workshop SMM4H: the 3rd social media mining for health applications workshop & shared task*, pages 13–16.

Lianwei Wu, Yuan Rao, Haolin Jin, Ambreen Nazir, and Ling Sun. 2019. Different absorption from the same sharing: Sifted multi-task learning for fake news detection. In *Proceedings of the 2019 Conference on Empirical Methods in Natural Language Processing and the 9th International Joint Conference on Natural Language Processing (EMNLP-IJCNLP)*, pages 4636–4645.

George-Eduard Zaharia, George-Alexandru Vlad, Dumitru-Clementin Cercel, Traian Rebedea, and Costin Chiru. 2020. Upb at semeval-2020 task 9: Identifying sentiment in code-mixed social media texts using transformers and multi-task learning. In *Proceedings of the Fourteenth Workshop on Semantic Evaluation*, pages 1322–1330.

Poorya Zaremoodi and Gholamreza Haffari. 2018. Neural machine translation for bilingually scarce scenarios: a deep multi-task learning approach. In *Proceedings of the 2018 Conference of the North American Chapter of the Association for Computational Linguistics: Human Language Technologies, Volume 1 (Long Papers)*, pages 1356–1365.

Pre-trained Transformer-based Classification and Span Detection Models for Social Media Health Applications

Yuting Guo and **Yao Ge**
Computer Science
Emory University
Atlanta GA 30322, USA
yuting.guo@emory.edu
yao.ge@emory.edu

Mohammed Al-Garadi and **Abeed Sarker**
Biomedical Informatics
Emory University
Atlanta GA 30322, USA
maalgar@emory.edu
abeed@dbmi.emory.edu

Abstract

This paper describes our approach for six classification tasks (Tasks 1a, 3a, 3b, 4 and 5) and one span detection task (Task 1b) from the Social Media Mining for Health (SMM4H) 2021 shared tasks. We developed two separate systems for classification and span detection, both based on pre-trained Transformer-based models. In addition, we applied oversampling and classifier ensembling in the classification tasks. The results of our submissions are over the median scores in all tasks except for Task 1a. Furthermore, our model achieved first place in Task 4 and obtained a 7% higher F_1-score than the median in Task 1b.

1 Introduction

Social media platforms such as Twitter have been widely used to share experiences and health information such as adverse drug effects (ADEs), thus attracting an increasing number of researchers to conduct health-related research using this data. However, because social media data consists of user-generated content that is noisy and written in informal language, health language processing with social media data is still challenging. To promote the use of social media for health information extraction and analysis, the Health Language Processing Lab of the University of Pennsylvania organized Social Media Mining for Health Applications (SMM4H) shared tasks. This year, the SMM4H shared tasks included 8 subtasks (Magge et al., 2021). Our team, the Sarker Lab at Emory University, participated in six classification tasks (*i.e.*, Task 1a, 3a, 3b, 4, and 5) and one span detection task (*i.e.*, Task 1b) of the SMM4H 2021 shared tasks. In recent years, Transformer-based models such as BERT (Devlin et al., 2019) and RoBERTa (Liu et al., 2019), whose advantage is modeling of long-range context semantics, revolutionised the field of NLP and achieved state-of-the-art results in different NLP tasks. Encouraged by those successes, we developed separate systems for classification and span detection both based on pre-trained Transformer-based models. We experimented with different Transformer-based model variants, and the model that achieved the best result on the validation set was selected as the final system. In addition, we performed undersampling and oversampling to address the problem of data imbalance and applied an ensemble technique in the classification tasks. The performances of our submissions are above the median F_1-scores in all tasks except for Task 1a. Furthermore, our model achieved first place in Task 4 and obtained a 7% higher F_1-score than the median in Task 1b.

2 Classification Tasks

2.1 Problem Definition and Datasets

We participated in six classification tasks including Task 1a: Classification of adverse effect mentions in English tweets; Task 3a and 3b: Classification of change in medication regimen in tweets and drug reviews from WebMD.com; Task 4: Classification of tweets self-reporting adverse pregnancy outcomes; Task 5: Classification of tweets self-reporting potential cases of COVID-19; and Task 6: Classification of COVID19 tweets containing symptoms. Further details about the data can be found in Magge et al. (2021). Among the six classification tasks, Task 6 was three-way classification and used micro-averaged F_1-score as the evaluation metric, while the remaining tasks were binary classification and used the F_1-score for the positive class for evaluation. For all tasks, we split the training data into a training set (`TRN`) and a validation set (`TRN_VAL`) with a 90/10 ratio, and evaluated the model on the validation set (`VAL`) released by the organizers.

2.2 Method

We used a uniform framework for all classification tasks, which consists of a Transformer-based en-

Proceedings of the Sixth Social Media Mining for Health Workshop 2021, pages 52–57
June 10, 2021.

Task	Task 1a	Task 3a	Task 3b	Task 4	Task 5	Task 6	Task 1a$_o$	Task 3a$_o$	Task 5$_o$
BT	64.5	59.6	88.4	89.3	71.6	98.4	67.2	60.4	71.9
CL	62.4	55.3	87.0	83.3	67.2	98.0	63.6	54.2	66.4
RBB	71.9	57.6	89.0	89.4	74.9	98.2	75.4	60.3	75.8
RBL	68.4	61.4	88.8	92.0	76.5	98.6	**78.6**	62.4	76.8
RBB+BT	66.7	62.5	89.1	91.2	79.2	98.6	73.9	64.7	77.0
RBB+RBL	69.1	66.7	89.4	92.7	80.3	98.8	75.2	65.6	79.2
RBB+CL	66.7	59.1	89.1	88.2	75.4	98.4	69.6	62.4	74.7
BT+RBL	68.5	65.7	89.5	**92.9**	78.7	**99.0**	76.5	**66.9**	78.3
BT+CL	66.7	59.4	88.7	87.4	72.8	98.4	67.9	61.7	74.1
RBL+CL	65.4	60.0	89.3	90.6	74.6	98.6	73.7	63.2	74.4
RBB+BT+RBL	67.3	64.9	89.4	92.5	**80.8**	98.8	74.3	66.4	79.1
RBB+BT+CL	67.3	61.5	89.0	89.7	74.8	98.8	67.9	63.8	76.7
RBB+RBL+CL	68.5	61.4	89.7	91.4	76.7	98.6	73.7	66.1	77.5
BT+RBL+CL	66.7	61.2	89.5	91.4	76.4	**99.0**	71.6	65.3	77.2
BT+CL+RBB+RBL	66.7	61.9	**89.8**	91.8	78.2	98.6	75.0	65.6	79.1

Table 1: F_1-scores of individual models and ensemble models on the validation (VAL) sets, where **Task*$_o$** denotes that the models are trained on the oversampled training (TRN) sets, and **ALL** denotes the ensemble of four individual models. The model that performed best on each task is highlighted in boldface.

coder, a pooling layer, a linear layer, and an output layer with Softmax activation. For each instance, the encoder converts each token into an embedding vector, and the pooling layer generates a document embedding by averaging the token embeddings. The document embedding is then fed into the linear layer and the output layer. The output is a probability vector with values between 0 and 1, which is used to compute a logistic loss during the training phase, and the class with the highest probability is chosen during the inference phase.

Encoder: Encouraged by the success of pre-trained Transformer-based language models, we experimented on four Transformer-based models pre-trained on different corpora–BERTweet (BT) (Nguyen et al., 2020) trained on English tweets, Bio_Clinical BERT (CL) (Alsentzer et al., 2019) on biomedical research papers and clinical notes, and RoBERTa$_{Base}$ (RBB) and RoBERTa$_{Large}$ (RBL) (Liu et al., 2019) on generic text such as English Wikipedia. We selected these models in order to investigate how the model size and the domain of pre-training data can benefit the performance on health-related tasks with social media data.

Preprocessing: To reduce the noise of tweets, we used the open source tool `preprocess-twitter` for data preprocessing.[1] The preprocessing includes lowercasing, normalization of numbers, usernames, urls, hashtags and text smileys, and adding extra marks for capital words, hashtags and repeated letters.

Oversampling: As described in Magge et al. (2021), the class distributions of Task 1a, Task 3a and Task 5 are imbalanced. To address the problem, we oversampled the minority class in the training set by picking samples at random with replacement using a Python toolkit called *imbalanced-learn*. The script is available on Github.[2] After oversampling, the new training sets included 28,942, 9644 and 9786 instances for Task 1a, Task 3a and Task 5, respectively.

Ensemble Modeling: In an attempt to improve performance over individual classifiers, we applied an ensemble technique to combine the results of different models. We averaged the outputs (*i.e.*, the probability vectors) of each model and selected the class with the highest value as the inference result.

2.3 Experiments and Results

We trained each model for 10 epochs, and the checkpoints that achieved the best performances on TRN_VAL were selected for evaluation. We experimented with two learning rates $\in \{2^{e-5}, 3^{e-5}\}$ and three different random initializations, meaning that there were six checkpoints in total for each model.[3] For each type of model, the median of the six checkpoints was used when we reported the results of individual models (*i.e.*, BT, CL, RBB, and RBL). For each ensemble model, all of the six checkpoints of the same type of model were

[1] https://nlp.stanford.edu/projects/glove/preprocess-twitter.rb

[2] https://gist.github.com/yguo0102/c72b5c0c353bea31bd7d72a15f6a0899

[3] Other hyper-parameters are fixed for all models. The batch size is 32, the weight decay is 0.1, and the warm-up ratio is 0.06.

used. Therefore, an ensemble model that combines k types of models consists of $6 \times k$ checkpoints.

	Task	Precision	Recall	F_1-Score
Ours	Task 1a	52.1	32.7	40.0
	Task 3a	72.1	63.5	68.0
	Task 3b	84.2	88.2	86.0
	Task 4	**93.9**	**92.2**	**93.0**
	Task 5	73.2	77.3	75.0
	Task 6	94.5	94.5	94.0
Median	Task 1a	50.5	40.9	44.0
	Task 4	91.8	92.3	92.5
	Task 5	73.9	74.4	74.5
	Task 6	93.2	93.2	93.0

Table 2: Our results and the median results on the evaluation sets of the classification tasks. The system that ranked first during the competition is highlighted in boldface.

Table 1 shows the results of individual models and ensemble models trained on the oversampled training sets. For each task, we submitted the model that performed best on the validation set, and the results of the test sets are shown in Table 2. The performances of our systems were above the median for each task except for Task 1a, and achieved first place on Task 4. For Task 3a and Task 3b, our system achieved 18% higher F_1-score on Task 3a and comparable result on Task 3b compared to the baseline model (Weissenbacher et al., 2020).[4]

2.4 Analysis

In general, for individual models, RoBERTa$_{Base}$ and RoBERTa$_{Large}$ performed better or comparable to BERTweet, and Bio_Clinical BERT underperformed on all tasks compared to the other models, which is consistent with our previous findings (Guo et al., 2020). Ensemble models outperformed individual models on all tasks except for Task 1a. We observed that for Task 1a, all models achieved high F_1-scores (around 97%) on the `TRN_VAL` set after training for 1 epoch, but the performance dropped by 25%-35% on the `VAL` set. Similarly, our F_1-score on the testing set of Task 1a is 40%, which is lower than that on the `VAL` set. Since the same trend is not present for other tasks, we hypothesized that the types of ADE in the training set and validation set of Task 1a may have low overlap.

To test our hypothesis, we counted the number of distinct ADE labels and normalized ADE labels using the data of Task 1b and Task 1c, shown in Table 3. Interestingly, the overlap percentage of normalized ADE labels is as high as 85.5%, and that of unnormalized ADE labels is much lower. This suggests that most types of ADE in the validation set are included in the training set but the ADE descriptions can vary widely. This result indicates that the gap between the performance on the training set and validation set may be attributed to the limited generalizability of pre-trained Transformer-based models to capture the semantic similarities between different expressions of the same ADE.

Type	Training	Validation	Overlap/percent
ADE	1127	85	35/41.2%
ADE$_n$	476	69	59/85.5%

Table 3: The number of the distinct ADE labels in the training set and validation set of Task 1, where **ADE$_n$** denotes the normalized ADE labels. The overlap percentage is computed based on the validation set.

3 Task 1b - ADE Span Detection

3.1 Problem Definition and Dataset

Task 1b aims at distinguishing adverse effect mentions from Non-ADE expressions and identifying the text spans of these adverse effect mentions. A tweet can have more than one ADE mention, and an ADE mention can be a sequence of words as well. The training set consists of 17,385 tweets annotated with 1713 ADE mentions for 1235 tweets, and the validation set consists of 915 tweets annotated with 87 ADE mentions for 65 tweets.

3.2 Method

We implemented several Transformer-based models including BioBERT (Lee et al., 2020), SciBERT (Beltagy et al., 2019), BERTweet (Nguyen et al., 2020) and two models of BERT (Devlin et al., 2019), and compared their performances.[5] BioBERT is specifically trained for biomedical text and widely used for the biomedical text-mining for NER. SciBERT is trained on more general domain data such as computer science text. BERTweet is a pre-trained language model for English Tweets. In addition, since the dataset is very imbalanced, we also performed undersampling to change the composition of the training set. Specifically, we randomly divided the training data with negative labels into 10 non-overlapping subsets, each of which

[4]Because we were the only participant for Task 3a and 3b, there is no median score available.

[5]For each of these 5 methods, we used the "cased" and "base" models if it is not specified.

has a slightly larger size (2000 tweets) compared to the positive data (the same 1235 positive tweets), and then 5 subsets were randomly selected for our experiment.

3.3 Experiments and Results

In our experiments, since the tweets are relatively short, we set the max sequence length to 128, batch size to 128 for BERT$_{Large}$ and 256 for other models. The learning rate was set to 5^{e-5}, and the epoch was set to 20 for all 5 models. The final submissions were evaluated in terms of precision, recall, and F$_1$-score by the official evaluation scripts provided by the organizers, for each ADE extracted where the spans overlap either entirely or partially. However, for the convenience of comparing the performance of the models during the experiments, we used Seqeval,[6] which is a Python framework for sequence labeling evaluation, to compare all methods on the validation set also by precision, recall, and F$_1$-score at the token level. Table 4 shows the performances for these 5 models.

Model	Precision	Recall	F$_1$-score
BioBERT	38.0	42.2	40.0
BERTweet	36.6	41.3	38.8
SciBERT	44.3	42.2	43.2
BERT$_{Base}$	**48.1**	39.1	43.1
BERT$_{Large}$	47.6	**46.9**	**47.3**

Table 4: The performances of models on validation set. The highest scores of precision, recall, and F$_1$-score have been highlighted in the table respectively.

From Table 4, it can be observed that BERT$_{Large}$ outperforms all other models with the highest recall and F$_1$-score. As a result, we chose BERT$_{Large}$ as the model used in the final submission. Finally, the result we received from the organizers was similar to the performance on the validation set, which is above the median. Although our recall is 17% worse than the median recall, our precision is 68.1 (+19%) and our F$_1$-score is 49.0 which is 7% higher than the median F$_1$-score.

3.4 Analysis

3.4.1 Comparison Between Models

We conducted the research on the learning efficiency and the performance over 20 epochs of each model, evaluating each time on the validation set. The results of precision, recall, and F$_1$-score for each epoch are shown in Figure 1.

[6]https://github.com/chakki-works/seqeval

These three plots show that the learning efficiency of BERT$_{Large}$ is very fast. When the epoch is 2, precision, recall and F$_1$-score for this model reach about 35%, while the scores of other models are only around 15% at this stage. In addition, as shown in the plots, the performance of BERT$_{Large}$ is consistently better than other models during training, which may benefit from its larger pre-training dataset. However, it is surprising to find that, unlike the curves of BioBERT, SciBERT and BERTweet, the curves of BERT$_{Base}$ model are relatively unstable, with some fluctuations.

3.4.2 Undersampling Experiments

Since BERT$_{Large}$ was the best model in our experiments, we separately finetuned BERT$_{Large}$ for 10 epochs on each of the 5 undersampled datasets, and compared the average scores for these 5 subsets with the performance scores obtained without undersampling. These results were also evaluated on the validation set at the token level. The results for undersampling are shown in Table 5. The averaged F$_1$-score for all the undersampled subsets is significantly lower than the best performance. Although we used all the positive data, it is possible that the drastic reduction in the amount of negative data and the total training data has had a very large impact on the results. Furthermore, randomly sampling the negative examples changes the prior distribution of the probability for the classifier. Due to time constraints associated with the shared task deadline, we were unable to try more advanced heuristics to select the negative examples for the undersampling, which is worth further exploring in future work.

Model	Precision	Recall	F$_1$-score
Subset-data1	22.1	62.5	32.7
Subset-data2	24.7	60.9	35.1
Subset-data3	22.7	63.6	33.3
Subset-data4	25.4	68.8	37.1
Subset-data5	23.0	64.1	33.9
AVG of Subset-data	**23.6**	**64.0**	**34.4**
With all training data	**47.6**	**46.9**	**47.3**

Table 5: Results when training on 5 undersampled datasets. Subset-data1 to Subset-data5 represent the 5 subsets that were randomly selected for experiment.

3.4.3 Performance Analysis

In order to conduct the research on the results we received from the organizers, we compared the annotated data for validation set provided by the or-

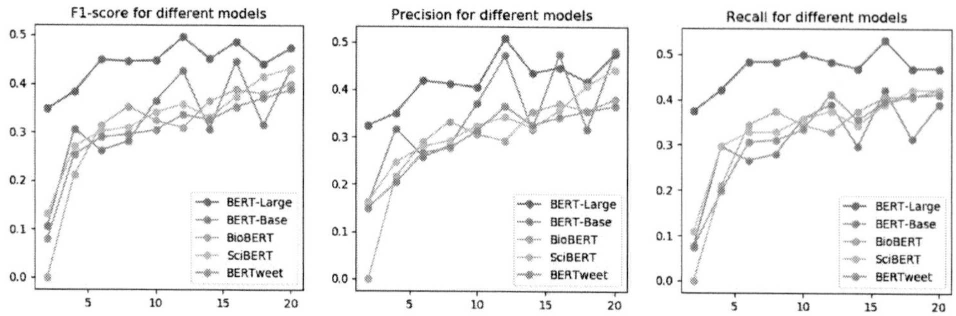

Figure 1: F_1-score, Precision and Recall for different models applied to the span detection task. The x-axis represents epoch.

ganizers with the results predicted by BERT$_{Large}$. This analysis revealed two primary causes why our model did not receive higher scores. Firstly, the number of true positives is relatively small. 87 annotations with label "ADE" were given in the validation set, but after the prediction, only 60 ADE mentions in the validation set (including true positive cases and false positive cases) were obtained. In these 60 ADE mentions, 23 cases which we only partially correctly predicted are also included, which means that many true ADEs were not detected (false negatives). Secondly, most of the ADE mentions predicted by our models, which are not annotated with label "ADE" in the validation set, did not appear for no reason, but actually have been annotated with label "ADE" in the training set. For example, "nosleep", which does not seem to have any ambiguity, is marked as ADE in one tweet, but not in another tweet, which might be due to the differences in the contexts in which they are mentioned. For example, in some tweets, "nosleep" appears in the tag "teamnosleep"; although it was predicted as ADE mention after being tokenized, it was not actually labeled as ADE by annotators.

4 Conclusion

In this work, we developed two systems based on pre-trained Transformer-based models for multiple health-related classification tasks and one span detection task for the SMM4H 2021 shared tasks. We experimented with different Transformer-based model variants as well as sampling strategies and applied an ensemble technique in the classification tasks. The results of our submissions are over the median F_1-scores in all tasks except for Task 1a. Furthermore, our model achieved first place in Task 4 and obtained a 7% higher F_1-score than the median in Task 1b. For future work, we will investigate methods to improve the generalizability of pre-trained Transformer-based models to deal with various health-related expressions in social media data.

References

Emily Alsentzer, John Murphy, William Boag, Wei-Hung Weng, Di Jindi, Tristan Naumann, and Matthew McDermott. 2019. Publicly Available Clinical BERT Embeddings. In *Proceedings of the 2nd Clinical Natural Language Processing Workshop*, pages 72–78, Minneapolis, Minnesota, USA. Association for Computational Linguistics.

Iz Beltagy, Kyle Lo, and Arman Cohan. 2019. SciBERT: A Pretrained Language Model for Scientific Text. volume 1903.

Jacob Devlin, Ming-Wei Chang, Kenton Lee, and Kristina Toutanova. 2019. BERT: Pre-training of Deep Bidirectional Transformers for Language Understanding. volume 1810.

Yuting Guo, Xiangjue Dong, Mohammed Ali Al-Garadi, Abeed Sarker, Cecile Paris, and Diego Mollá Aliod. 2020. Benchmarking of Transformer-Based Pre-Trained Models on Social Media Text Classification Datasets. In *Proceedings of the The 18th Annual Workshop of the Australasian Language Technology Association*, pages 86–91, Virtual Workshop. Australasian Language Technology Association.

Jinhyuk Lee, Wonjin Yoon, Sungdong Kim, Donghyeon Kim, Sunkyu Kim, Chan Ho So, and Jaewoo Kang. 2020. BioBERT: a pre-trained biomedical language representation model for biomedical text mining. In *Bioinformatics*, volume 36, page 985–989.

Yinhan Liu, Myle Ott, Naman Goyal, Jingfei Du, Mandar Joshi, Danqi Chen, Omer Levy, Mike Lewis, Luke Zettlemoyer, and Veselin Stoyanov. 2019. RoBERTa: A Robustly Optimized BERT Pretraining Approach. *arXiv*, 1907(11692).

Arjun Magge, Ari Klein, Ivan Flores, Ilseyar Alimova, Mohammed Ali Al-garadi, Antonio Miranda-Escalada, Zulfat Miftahutdinov, Eulàlia Farré-Maduell, Salvador Lima López, Juan M Banda, Karen O'Connor, Abeed Sarker, Elena Tutubalina, Martin Krallinger, Davy Weissenbacher, and Graciela Gonzalez-Hernandez. 2021. Overview of the sixth social media mining for health applications (# SMM4H) shared tasks at NAACL 2021. In *Proceedings of the Sixth Social Media Mining for Health Applications Workshop & Shared Task*.

Dat Quoc Nguyen, Thanh Vu, and Anh Tuan Nguyen. 2020. BERTweet: A Pre-trained Language Model for English Tweets. In *Proceedings of the 2020 Conference on Empirical Methods in Natural Language Processing: System Demonstrations*, pages 9–14.

Davy Weissenbacher, Suyu Ge, Ari Klein, Karen O'Connor, Robert Gross, Sean Hennessy, and Graciela Gonzalez-Hernandez. 2020. Active Neural Networks to Detect Mentions of Changes to Medication Treatment in Social Media. *medRxiv*.

BERT Goes Brrr: A Venture Towards the Lesser Error in Classifying Medical Self-Reporters on Twitter

Alham Fikri Aji[∗•] **Haryo Akbarianto Wibowo**[∗]
Made Nindyatama Nityasya[∗] **Radityo Eko Prasojo**[∗⋆] **Tirana Noor Fatyanosa**[∗◦]
[∗] Kata.ai Research Team, Jakarta, Indonesia
[•] School of Informatics, University of Edinburgh
[⋆] Faculty of Computer Science, Universitas Indonesia
[◦] Graduate School of Science and Technology, Kumamoto University
{aji,haryo,made,ridho,tirana.fatyanosa}@kata.ai
fatyanosa@dbms.cs.kumamoto-u.ac.jp

Abstract

This paper describes Kata.ai's submission for the Social Media Mining for Health (SMM4H) 2021 shared task. We participated in three tasks: classifying adverse drug effect, COVID-19 self-report, and COVID-19 symptoms. Our system is based on BERT model pre-trained on the domain-specific text. In addition, we perform data cleaning and augmentation, as well as hyperparameter optimization and model ensemble to further boost the BERT performance. We achieved the first rank in both classifying adverse drug effects and COVID-19 self-report tasks.

1 Introduction

Over the years, social media has been used as a massive data source to monitor health-related issues (Weissenbacher et al., 2018, 2019; Klein et al., 2020), such as flu trends (Achrekar et al., 2011; Paul and Dredze, 2012), adverse drug effects (Cocos et al., 2017; Pierce et al., 2017), or viral disease outbreak such as the COVID-19 (Sarker et al., 2020; Klein et al., 2021). In general, leveraging massive self-reported data is considered useful for supplementing the otherwise long and costly process of clinical trials in obtaining a more comprehensive picture of the issue in hand.

Nevertheless, analyzing text data from social media is challenging due to its noisy nature, which stems from the prevalence of linguistic errors and typos. In this work, we leverage BERT (Devlin et al., 2018) to handle noisy text through domain-specific pre-training, data cleaning, augmentation, hyperparameter optimization, and model ensemble. With this training pipeline, we achieved the best performance in Social Media Mining for Health (SMM4H) 2021 shared task (Magge et al., 2021) for classifying Adverse Drug Effect and COVID-19 self-report from Twitter text.

Text	Label
How is it that Vyvanse gives me dry mouth, but I still produce this much saliva in my sleep?	ADE
I need Temazepam and alprazolam.... Is there any doctor can prescribe for me?? :/	NoADE

(a) Task 1a : Classification of adverse drug effect (ADE) mentions in English tweets

Text	Label
This girl in my class really had the coronavirus, I'm booking an appointment with my doctor for a check up	1
Read someone on facebook say she hopes the coronavirus doesn't come with the goods she ordered online. Either way, you're quarantined from my facebook, you racist bitch!	0

(b) Task 5 : Classification of tweets self-reporting potential cases of COVID-19

Text	Label
Maybe they've been asked too early. I had a total loss of smell and taste in week 3. In week 1 I only had phantom smells and that's when you test positive.	Self
My brother came home from Paris with a sore throat and a fever and I know he gave me coronavirus. I KNOW IT.	Nonpersonal
Months after Covid-19 infection, patients report breathing difficulty and fatigue https://t.co/H3wcVLxL6y	Lit-News

(c) Task 6 : Classification of COVID-19 tweets containing symptoms

Table 1: Task data examples.

2 BERT Goes Brrr

We participated in 3 classification tasks: Task 1a to classify the adverse drug effect (ADE), Task 5 to classify COVID-19 potential case, and Task 6 to classify COVID-19 symptoms. The distribution of

Proceedings of the Sixth Social Media Mining for Health Workshop 2021, pages 58–64
June 10, 2021. ©2021 Association for Computational Linguistics

Task 1			Task 5			Task 6		
Label	Train	Valid	Label	Train	Valid	Label	Train	Valid
ADE	1231	65	0	5439	594	Lit-News_mentions	4277	247
NoADE	16113	848	1	1026	122	Nonpersonal_reports	3442	180
						Self_reports	1348	73
All	17344	913	All	6465	717	All	9067	500

Table 2: Distribution of datasets.

the datasets are given in Table 2. All tasks' text data are taken from Twitter, with some examples shown in Table 1. More detailed information about the dataset can be found in (Klein et al., 2021; Magge et al., 2021).

We used BERT for all three tasks, implemented with the Huggingface toolkit (Wolf et al., 2020). For each task, we started off by fine-tuning the off-the-shelf BERT-base (Devlin et al., 2018), which resulted in a fairly good performance (Table 3). Then, we improved by using domain-specific BERT instead, then by performing data cleaning, data augmentation, hyperparameter optimization, and finally model ensembling. Table 3 shows the F1-Score improvement by incorporating each of those techniques. Detailed experiments for each technique are in Section 3. Note that some techniques are not used in certain tasks, specified by the dash symbol on the table.

Among the 3 tasks, we achieved the best score for Task 1a and Task 5. Our standing for Task 6 by the time of this paper submission is currently unknown. Still, our performance on Task 6 is above the median, as seen in Table 4. We note that our Task 1a performance on the test set drops significantly compared to its performance on the valid set, indicating overfitting on the valid set. Unfortunately, further analysis on the test set was not feasible since the labels are not provided.

3 Improving BERT

In this section, we dissect each technique we introduce to our submission model.

3.1 Baseline Model

BERT (Devlin et al., 2018) is a pretrained language model based on the Transformer (Vaswani et al., 2017). It is, alongside its many variants, the current state-of-the-art for many NLP applications. It also dominates the previous year's SMM4H shared task and comes out as the winning system (Klein et al., 2020; Weissenbacher et al., 2019).

There are many BERT pre-trained models. To have a good starting point, we explored several pre-trained models. First, we compared general BERT models such as DistilBERT, ALBERT, BERT-base,[1] and BERT-large.[2] Then, knowing that our datasets are tweets that potentially contain medical terms, we explored some domain-specific models: Bio-ClinicalBERT[3] which is trained on biomedical and clinical text (Alsentzer et al., 2019), BERTweet[4] which is trained on English tweets (Nguyen et al., 2020), and BERTweet-Covid19[5] which is built by continuing the pre-trained BERTweet using English tweets related to COVID-19 (Nguyen et al., 2020). We found that BERTweet-Covid19 gives the best result even in the non-COVID-19 related data like Task 1's ADE (see Table 5).

We also considered another COVID-19 tweets pretrained model, that is COVID-Twitter-BERT (CT-BERT)[6] (Müller et al., 2020). It is based on the BERT-large model, while the BERTweet-Covid19 is a BERT-base model. We found that fine-tuning on this model using the recommended hyperparameters is relatively unstable compared to the BERTweet-Covid19 model, though it does outperform it occasionally. As such, we used this model in the later steps, that is, only with hyperparameter optimization and ensembling.

3.2 Data Cleaning

We focused on eliminating tokens that are potential sources of bias. We found that masking Twitter handles, URLs, emails, phone numbers, and money yields the best results. In our experiments, masking all numerical tokens produces worse results.

Furthermore, we also performed a routine HTML tag cleanup, as well as hashtag expan-

[1] https://huggingface.co/bert-base-uncased
[2] https://huggingface.co/bert-large-uncased
[3] https://huggingface.co/emilyalsentzer/Bio_ClinicalBERT
[4] https://huggingface.co/vinai/bertweet-base
[5] https://huggingface.co/vinai/bertweet-covid19-base-uncased
[6] https://huggingface.co/digitalepidemiologylab/covid-twitter-bert

Method	Valid F1-Score		
	Task 1a	Task 5	Task 6
BERT-base model	70.87	73.03	98.27
+ Domain-specific BERT	79.14	74.60	98.55
+ Data Cleaning	82.93	77.64	98.87
+ Data Augmentation	–	80.31	–
+ Hyperparameter Optimization	84.30	82.20	–
+ Model Ensembling (submitted system)	87.80	86.27	98.90

Table 3: Our system performance on valid set.

Task	Our Performance			Median Performance			Standing
	F1-Score	Precision	Recall	F1-Score	Precision	Recall	
Task 1a	0.54	0.603	0.489	0.44	0.505	0.409	1st place
Task 5	0.79	0.781	0.789	0.74	0.739	0.743	1st place
Task 6	0.94	0.944	0.944	0.93	0.932	0.932	-

Table 4: Our submitted system performance on test set, compared with the median performance.

Method	Task 1a	Task 5	Task 6
BERT-base	70.87	73.03	98.27
BERT-large	77.78	73.60	98.39
Bio-ClinicalBERT	68.97	68.57	97.33
BERTweet	75.76	71.09	**98.55**
BERTweet-Covid19	**79.14**	**74.60**	**98.55**

Table 5: Baseline on valid set. Task 1a: F1-Score for the ADE class. Task 5: F1-Score for the "potential case" class. Task 6: Macro F1-Score for all classes.

sion (e.g., "#SaveTheEarth" becomes "save the earth"). To this end, we leveraged Ekphrasis (Baziotis et al., 2017) tokenization and masking pipeline. Finally, we performed emoji codification (e.g. into `:thumbsup:`, `:red_heart:`, etc.) using the python emoji package.[7] The emoji codes are treated as special tokens, following the configuration of our chosen base models (Nguyen et al., 2020; Müller et al., 2020).

Below are some data cleaning attempts that did not improve our final model performance.

1. We handpicked some relevant Twitter handles to keep unmasked (such as @WHO). We also tried to pick top-*n* most frequent handles to stay unmasked. Both did not yield better results.
2. We crawled the URLs to get their titles. Using a keyword-based extraction, we determine whether the title is relevant to COVID-19, and if so, we append the title to the end of the tweet. This did not improve the performance of our models.
3. We tried to fix grammatical and typography informality (such as the use of contraction) using Ekphrasis's toolkit, which is based on Norvig's spell checker algorithm. This does not provide better results, not even when using BERT-base or BERT-large.

[7] https://pypi.org/project/emoji/

3.3 Data Augmentation

The provided training data is imbalanced: the number of positive class data is significantly less (Table 2). Therefore, we tried 2 approaches to deal with this issue, namely data oversampling and class weighting. In data oversampling, we duplicate the minority class training data. On the other hand, class weighting simply increases the gradient weight of the minority class.

Additional training data, including the synthetic one, has been shown to improve the model performance (Wei and Zou, 2019; Ma, 2019). Hence, we also explored augmentation data by paraphrasing the training. We create paraphrases by using round-trip translation (Mallinson et al., 2017): our English dataset is translated into another pivot language, then translated back into English. We've tried different pivot languages as well as different translation engines. Based on our manual judgement, using Google Translate and German as the pivot provides the best paraphrase.

Method	Data size	F1
BERTweet-Covid19 (Bc19)	6.5k	74.60
Bc19 + Class-weight	6.5k	75.20
Bc19 + Oversampling	10.5k	76.74
Bc19 + Paraphrase Aug.	12.9k	76.45
Bc19 + Class-weight + Paraphrase Aug.	12.9k	76.49
Bc19 + Oversampling + Paraphrase Aug.	21.1k	**77.65**

Table 6: Task 5 result on data augmentation

Experimental results on data augmentation and data balancing can be seen in Table 6. Our result shows that oversampling is better than class-weighting for dealing with imbalanced training data. Orthogonally, data augmentation can also improve performance. The combination of both data oversampling and data augmentation can increase performance even higher. However, it should be noted that the size of the training data has also increased significantly.

Note that our baseline in this experiment is BERTweet-Covid19 without data cleaning. On uncleaned raw input, we achieved F1-Score of 77.65, as shown in Table 6. However, applying oversampling + paraphrase augmentation on cleaned data can further improve the F1-Score to 80.31.

Interestingly, Task 1a does not benefit from data augmentation or data balancing. Furthermore, adding extra training data from past years' training set does not help as well. Therefore, we only apply data augmentation for Task 5.

3.4 Hyperparameter Optimization

Nowadays, it is common knowledge that optimizing hyperparameter can improve the performance of machine learning algorithms (Kaur et al., 2020; Yang and Shami, 2020; Fatyanosa and Aritsugi, 2020). Current research on the transformer (Murray et al., 2019; Zhang and Duh, 2020) also moving towards hyperparameter optimization (HPO) as the transformer models are susceptible on the chosen hyperparameters (Murray et al., 2019).

The purpose of this section is to determine the best hyperparameter combination of the baseline model for Task 1a and Task 5. We did not optimize the model for Task 6 as the results were already good.

HPO is a time-consuming task. Therefore, performing manual HPO would be inefficient, and it is advisable to utilize automatic optimization. There are several well-known automatic HPO approaches. In this paper, we only use bayesian HPO, specifically, the Tree-structured Parzen Estimator (TPE).

TPE selects the next possible combination of hyperparameters by building probabilistic models. To simplify the search process of the best hyperparameter combination, we employ the Hyperopt (Bergstra et al., 2013) package. As stated in Section 3.1, we also explored a stable and better hyperparameter configuration for Covid-Twitter-BERT.

Table 7 shows all the optimized hyperparameters and their ranges and values. The range for BS was selected following the capabilities of our GPU. We tried two optimizers: AdamW (Loshchilov and Hutter, 2017) and AdaBelief (Zhuang et al., 2020). The ranges for LR, EPS, and WD were selected based on recommendation from (Zhuang et al., 2020).

Hyper-parameter	Definition	Range/Value
BS	Batch size	Min: 8, Max: 32
LR	Learning rate	Min: 1e-6, Max: 1e-4
OP	Optimizer	['AdamW', 'AdaBelief']
EPS	Epsilon	Min: 1e-16, Max: 1e-8
WD	Weight Decay	Min: 0, Max: 1e-2

Table 7: Hyperparameter Range

We set the same random seed to 1 for our baseline. In HPO experiments, we tried to open the possibility of a better model by randomizing the seeds. This assumption is based on several studies suggesting that random seeds influence machine learning algorithms (Madhyastha and Jain, 2019; Risch and Krestel, 2020). It is important to note that the random seeds were not tuned; instead, they were generated randomly in each iteration of the TPE.

As predicted, HPO indeed increase the F1-Score for Task 1a and Task 5 when training the baseline model. After HPO, the results for Task 1a increased by 1.65% and Task 5 increased by 2.35% as shown in Table 3.

The HPO implementations for the two tasks were executed in the same search space and the same total number of iterations (100 iterations). The visual comparison of the results is illustrated in Figure 1. It shows that the optimal solution for Task 1a is obtained after 87 iterations. Meanwhile, Task 5 only needs 14 iterations. Although the faster discovery of the best combination is preferable in terms of execution time, this scenario can also mean that the algorithm is stuck in local optima.

In terms of execution time, an average of 21 min and 41 min were needed to finish an iteration for Task 1a and Task 5, respectively. Note that the execution time may vary depending on the model and the GPU. The HPO was implemented on NVIDIA Tesla V100 GPU.

Owing to the validation data optimization, it is predictable that HPO bias towards the validation set. Consequently, the model shown strong overfitting, specifically for Task 1a, where the results obtained are very far from the baseline results. The next step to combat the overfitting is to employ ensemble methods.

Task	Hyperparameter					Model	Best F1-Score
	BS	LR	OP	EPS	WD		
Task 1a	11	1.55E-05	AdamW	2.08E-09	0.002085	vinai/bertweet-covid19-base-cased	84.30
Task 5	19	5.21E-05	AdaBelief	5.10E-09	0.000421	digitalepidemiologylab/covid-twitter-bert	82.20

Table 8: HPO results

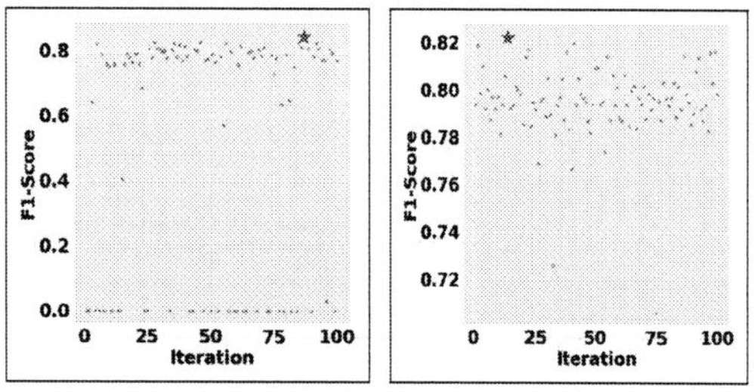

(a) Task 1a (b) Task 5

Figure 1: Visual comparison of HPO

3.5 Model Ensemble

Motivated by some past successful results (Chen et al., 2019; Casola and Lavelli, 2020), we ensembled some trained models on Task 1a and Task 5, which are picked from the best performing HPO models. In the implementation of the ensemble technique, to predict the label of an instance, we summed all of the chosen models' probability score and took the highest score as the label.

Typically, a model ensemble considers all of the chosen models. However, our experiments showed that this configuration does not produce the best results for Task 1a (Table 9). We then proceeded to perform an exhaustive search for every possible combinations (that is, the power set) of the chosen models.

Method	Task 1a	Task 5
Best HPO result	84.30 (1 model)	82.20 (1 model)
All Ensemble	82.71 (15 models)	83.40 (10 models)
Best Ensemble	**87.81** (5 models)	**86.28** (5 models)
Top-5 Ensemble	83.58 (5 models)	82.03 (5 models)

Table 9: Result of the model ensemble. **All Ensemble** ensembles all handpicked models. **Best Ensemble** ensemble the subset model of all handpicked models. **Top-5 Ensemble** is the top five best model ensemble result. The "**n** models" represents number of models used to produce the result.

As shown in Table 9, we found the best ensemble involves a subset of five models for both Task 1a and Task 5. There is also a significant gap between the performance of the best subset ensemble with the full model ensemble for both tasks. Regarding Task 1a's "All Ensemble" worse performance, we hypothesize that there might be some "noisy" models among the chosen ones. While our exhaustive search may alleviate this problem, it takes a lot of time that also increases exponentially with respect to the number of chosen models. We leave optimizing this process as future work.

Interestingly, simply choosing the best-performing models does not produce the best ensembled model. As shown in Table 9, model ensemble of the top-5 best F1 ("Top-5 Ensemble") performs worse than the "Best Ensemble". In fact, top-5 ensemble performed worse than a single non-ensembled model from the best HPO result.

4 Conclusion

We describe our team submission for Social Media Mining for Health Applications shared task 2021. Our system achieved the best performance for classifying Adverse Effect mentions and self-reporting potential cases of COVID-19 in English tweets.

Our system is based on BERT model. We observe improvement over the off-the-shelf BERT-base from using domain-specific BERT, rigorous data cleaning, data augmentation, hyperparameter optimization, and model ensembling. Among those techniques, we find that domain-specific BERT, data cleaning, and model ensembling improve the performance on all tasks, whereas data augmentation and hyperparameter optimization are more situational.

Overall, we obtain 17% and 13% improvement on Task 1a and Task 5 respectively (Table 3). On Task 6, we only obtain 0.6% improvement. This is because we did not perform data augmentation and hyperparameter optimization on this dataset, and because the base model already returns a high score of 98.27. We argue that these training pipelines can be used to improve the performance of general text classification tasks.

5 Acknowledgements

We thank Masayoshi Aritsugi (Kumamoto University) for comments on the manuscript.

References

Harshavardhan Achrekar, Avinash Gandhe, Ross Lazarus, Ssu-Hsin Yu, and Benyuan Liu. 2011. Predicting flu trends using twitter data. In *2011 IEEE conference on computer communications workshops (INFOCOM WKSHPS)*, pages 702–707. IEEE.

Emily Alsentzer, John Murphy, William Boag, Wei-Hung Weng, Di Jindi, Tristan Naumann, and Matthew McDermott. 2019. Publicly available clinical BERT embeddings. In *Proceedings of the 2nd Clinical Natural Language Processing Workshop*, pages 72–78, Minneapolis, Minnesota, USA. Association for Computational Linguistics.

Christos Baziotis, Nikos Pelekis, and Christos Doulkeridis. 2017. DataStories at SemEval-2017 task 4: Deep LSTM with attention for message-level and topic-based sentiment analysis. In *Proceedings of the 11th International Workshop on Semantic Evaluation (SemEval-2017)*, pages 747–754, Vancouver, Canada. Association for Computational Linguistics.

J. Bergstra, D. Yamins, and D. D. Cox. 2013. Making a science of model search: Hyperparameter optimization in hundreds of dimensions for vision architectures. In *Proceedings of the 30th International Conference on International Conference on Machine Learning - Volume 28*, ICML'13, page I–115–I–123. JMLR.org.

Silvia Casola and Alberto Lavelli. 2020. Fbk@ smm4h2020: Roberta for detecting medications on twitter. In *Proceedings of the Fifth Social Media Mining for Health Applications Workshop & Shared Task*, pages 101–103.

Shuai Chen, Yuanhang Huang, Xiaowei Huang, Haoming Qin, Jun Yan, and Buzhou Tang. 2019. Hitsz-icrc: a report for smm4h shared task 2019-automatic classification and extraction of adverse effect mentions in tweets. In *Proceedings of the fourth social media mining for health applications (# SMM4H) workshop & shared task*, pages 47–51.

Anne Cocos, Alexander G Fiks, and Aaron J Masino. 2017. Deep learning for pharmacovigilance: recurrent neural network architectures for labeling adverse drug reactions in twitter posts. *Journal of the American Medical Informatics Association*, 24(4):813–821.

Jacob Devlin, Ming-Wei Chang, Kenton Lee, and Kristina Toutanova. 2018. BERT: pre-training of deep bidirectional transformers for language understanding. *CoRR*, abs/1810.04805.

Tirana Noor Fatyanosa and Masayoshi Aritsugi. 2020. Effects of the Number of Hyperparameters on the Performance of GA-CNN. In *2020 IEEE/ACM International Conference on Big Data Computing, Applications and Technologies (BDCAT)*, pages 144–153. IEEE.

Sukhpal Kaur, Himanshu Aggarwal, and Rinkle Rani. 2020. Hyper-parameter optimization of deep learning model for prediction of Parkinson's disease. *Machine Vision and Applications*, 31(5):32.

Ari Klein, Ilseyar Alimova, Ivan Flores, Arjun Magge, Zulfat Miftahutdinov, Anne-Lyse Minard, Karen O'connor, Abeed Sarker, Elena Tutubalina, Davy Weissenbacher, et al. 2020. Overview of the fifth social media mining for health applications (# smm4h) shared tasks at coling 2020. In *Proceedings of the Fifth Social Media Mining for Health Applications Workshop & Shared Task*, pages 27–36.

Ari Z Klein, Arjun Magge, Karen O'Connor, Jesus Ivan Flores Amaro, Davy Weissenbacher, and Graciela Gonzalez Hernandez. 2021. Toward using twitter for tracking covid-19: A natural language processing pipeline and exploratory data set. *Journal of medical Internet research*, 23(1):e25314.

Ilya Loshchilov and Frank Hutter. 2017. Decoupled weight decay regularization. *arXiv preprint arXiv:1711.05101*.

Edward Ma. 2019. Nlp augmentation. https://github.com/makcedward/nlpaug.

Pranava Madhyastha and Rishabh Jain. 2019. On model stability as a function of random seed. *arXiv*, pages 929–939.

Arjun Magge, Ari Klein, Ivan Flores, Ilseyar Alimova, Mohammed Ali Al-garadi, Antonio Miranda-Escalada, Zulfat Miftahutdinov, Eulàlia Farré-Maduell, Salvador Lima López, Juan M Banda, Karen O'Connor, Abeed Sarker, Elena Tutubalina, Martin Krallinger, Davy Weissenbacher, and Graciela Gonzalez-Hernandez. 2021. Overview of the sixth social media mining for health applications (# smm4h) shared tasks at naacl 2021. In *Proceedings of the Sixth Social Media Mining for Health Applications Workshop & Shared Task*.

Jonathan Mallinson, Rico Sennrich, and Mirella Lapata. 2017. Paraphrasing revisited with neural machine translation. In *Proceedings of the 15th Conference of the European Chapter of the Association for Computational Linguistics: Volume 1, Long Papers*, pages 881–893.

Martin Müller, Marcel Salathé, and Per E Kummervold. 2020. Covid-twitter-bert: A natural language processing model to analyse covid-19 content on twitter. *arXiv preprint arXiv:2005.07503*.

Kenton Murray, Jeffery Kinnison, Toan Q. Nguyen, Walter Scheirer, and David Chiang. 2019. Auto-sizing the transformer network: Improving speed, efficiency, and performance for low-resource machine translation. *arXiv*, (Wngt):231–240.

Dat Quoc Nguyen, Thanh Vu, and Anh Tuan Nguyen. 2020. BERTweet: A pre-trained language model for English Tweets. In *Proceedings of the 2020 Conference on Empirical Methods in Natural Language Processing: System Demonstrations*.

Michael J Paul and Mark Dredze. 2012. A model for mining public health topics from twitter. *Health*, 11(16-16):1.

Carrie E Pierce, Khaled Bouri, Carol Pamer, Scott Proestel, Harold W Rodriguez, Hoa Van Le, Clark C Freifeld, John S Brownstein, Mark Walderhaug, I Ralph Edwards, et al. 2017. Evaluation of facebook and twitter monitoring to detect safety signals for medical products: an analysis of recent fda safety alerts. *Drug safety*, 40(4):317–331.

Julian Risch and Ralf Krestel. 2020. Bagging {BERT} Models for Robust Aggression Identification. *Proceedings of the Second Workshop on Trolling, Aggression and Cyberbullying*, (May):55–61.

Abeed Sarker, Sahithi Lakamana, Whitney Hogg-Bremer, Angel Xie, Mohammed Ali Al-Garadi, and Yuan-Chi Yang. 2020. Self-reported covid-19 symptoms on twitter: an analysis and a research resource. *Journal of the American Medical Informatics Association*, 27(8):1310–1315.

Ashish Vaswani, Noam Shazeer, Niki Parmar, Jakob Uszkoreit, Llion Jones, Aidan N Gomez, Lukasz Kaiser, and Illia Polosukhin. 2017. Attention is all you need. *arXiv preprint arXiv:1706.03762*.

Jason Wei and Kai Zou. 2019. EDA: Easy data augmentation techniques for boosting performance on text classification tasks. In *Proceedings of the 2019 Conference on Empirical Methods in Natural Language Processing and the 9th International Joint Conference on Natural Language Processing (EMNLP-IJCNLP)*, pages 6382–6388, Hong Kong, China. Association for Computational Linguistics.

Davy Weissenbacher, Abeed Sarker, Arjun Magge, Ashlynn Daughton, Karen O'Connor, Michael Paul, and Graciela Gonzalez. 2019. Overview of the fourth social media mining for health (smm4h) shared tasks at acl 2019. In *Proceedings of the fourth social media mining for health applications (# SMM4H) workshop & shared task*, pages 21–30.

Davy Weissenbacher, Abeed Sarker, Michael Paul, and Graciela Gonzalez. 2018. Overview of the third social media mining for health (smm4h) shared tasks at emnlp 2018. In *Proceedings of the 2018 EMNLP workshop SMM4H: the 3rd social media mining for health applications workshop & shared task*, pages 13–16.

Thomas Wolf, Lysandre Debut, Victor Sanh, Julien Chaumond, Clement Delangue, Anthony Moi, Pierric Cistac, Tim Rault, Remi Louf, Morgan Funtowicz, Joe Davison, Sam Shleifer, Patrick von Platen, Clara Ma, Yacine Jernite, Julien Plu, Canwen Xu, Teven Le Scao, Sylvain Gugger, Mariama Drame, Quentin Lhoest, and Alexander Rush. 2020. Transformers: State-of-the-art natural language processing. In *Proceedings of the 2020 Conference on Empirical Methods in Natural Language Processing: System Demonstrations*, pages 38–45, Online. Association for Computational Linguistics.

Li Yang and Abdallah Shami. 2020. On hyperparameter optimization of machine learning algorithms: Theory and practice. *Neurocomputing*, 415:295–316.

Xuan Zhang and Kevin Duh. 2020. Reproducible and Efficient Benchmarks for Hyperparameter Optimization of Neural Machine Translation Systems. *Transactions of the Association for Computational Linguistics*, 8:393–408.

Juntang Zhuang, Tommy Tang, Yifan Ding, Sekhar Tatikonda, Nicha Dvornek, Xenophon Papademetris, and James Duncan. 2020. Adabelief optimizer: Adapting stepsizes by the belief in observed gradients. *Conference on Neural Information Processing Systems*.

UACH at SMM4H: a BERT based approach for classification of COVID-19 Twitter posts

Alberto Valdés Chávez
Jesús Roberto López Santillán
Facultad de Ingeniería
Universidad Autónoma de Chihuahua
Chihuahua, Chih., Mexico
valdeschaveza@gmail.com
jrlopez@uach.mx

Manuel Montes-y-Gómez
Department of Computational Sciences
INAOE
Sta. Ma. Tonantzintla, Puebla, Mexico
mmontesg@inaoep.mx

Abstract

This work describes the participation of the Universidad Autónoma de Chihuahua team at the Social Media Mining for Health Applications (SMM4H) 2021 shared task. Our team participated in Tasks 5 and 6, both focused on the automatic classification of tweets related to COVID-19. Task 5 considered a binary classification problem, aiming to identify self-reporting tweets of potential cases of COVID-19. On the other hand, Task 6 goal was to classify tweets containing COVID-19 symptoms. For both tasks we used models based on bidirectional encoder representations from transformers (BERT). Our objective was to determine whether a model trained on a corpus from the domain of interest could outperformed one trained on a much larger general domain corpus. Our F1 results were encouraging, 0.77 and 0.95 for Tasks 5 and 6 respectively, having achieved the highest score among all the participants in the latter.

1 Introduction

The Social Media Mining for Health Applications (SMM4H) 2021 shared task aimed to address the challenges presented in Natural Language Processing (NLP) applied to text obtained from social net works, specifically Twitter, to gain medical insights (Magge et al., 2021). This year's SMM4H proposed 8 different problems that involved classification and Named Entity Recognition (NER) tasks. Our team focused on Tasks 5 and 6, both dealing with *classification of COVID-19 related tweets* in different situations. We decided to approach this problem with transformer-based models (Vaswani et al., 2017), since they are considered state-of-the-art in many NLP applications. Also, we hypothesized that a model trained on domain specific texts could performed better than one trained in a much larger but general-domain corpus. To test our hypothesis, we implemented two models that share the same architecture but were trained on different data sets; on the one hand, the large uncased version of BERT (BERT-Large) (Devlin et al., 2018), which is a model pretrained on a very large corpus (Wikipedia), and, on the other hand, CT-BERT that is a model based on BERT-Large but pretrained on a smaller corpus of COVID-19 related tweets (Müller et al., 2020).

2 Tasks Description

2.1 Task 5: Classification of tweets self-reporting potential cases of COVID-19

This is a binary classification task that involves distinguishing tweets of potential cases of COVID-19 (including situations that pose high risk of contagion) annotated as "1", from those that do not represent danger (annotated as "0"). Next, we show a tweet example for each class (Klein et al., 2021).

I think I have the coronavirus I've been coughing nonstop all day and I feel really warm **Label: "1"**

With coronavirus we certainly need more doctors and surgeons and nurses and sonographs and radiologists, let them in, quick! **Label: "0"**

Table 1 shows the labels distribution over the training, validation and test sets, as given by the organizers. NA denotes that the corresponding number is currently unknown.

Dataset	**"1"**	**"0"**	**#**
Training	1,026	5,439	6,465
Validation	122	594	716
Test	NA	NA	10,000

Table 1: Distribution of labels over the training, validation and test sets for Task 5.

Proceedings of the Sixth Social Media Mining for Health Workshop 2021, pages 65–68
June 10, 2021. ©2021 Association for Computational Linguistics

2.2 Task 6: Classification of COVID-19 tweets containing symptoms

This task is a three-way classification problem where the target classes are self_reports, non-personal_reports and literature/news_mentions. Self reports are personal mentions where the user describes his/her own symptoms. Nonpersonal reports are tweets where the user describes symptoms that other people experience. In addition, literature/news mentions are tweets coming from news articles or other sources that describe medical symptoms. Next, we show a tweet example for each class; then, Table 2 shows the labels distribution over the dataset.

In a study done in Milan, Italy, 402 Covid-19 patients were surveyed after being discharged. 28% showed symptoms of PTSD, 31% suffered from depression, 40% had insomnia, amp; 42% had anxiety. Overall, 56% of participants manifested at least one mental disorder following the disease. **Label: "Lit-News_mentions"**

@mention My wife takes 3 subway and a bus one way to reach her downtown office. She started having erratic fever and slight cough in past 3 days. She also travelled from India on 18th January via Germany and London. Does is qualify for a covid-19 test? **Label: "Nonpersonal_reports"**

Agreed! My covid19 was considered mid-level I wouldn't wish what I went through on my worst enemy. 1st symptoms March 19th - STILL RECOVERING!!! #longtailcovid **Label: "Self_reports"**

Dataset	LitNews	NonP	Self R	#
Training	4,277	3,442	1,348	9,067
Validation	247	180	73	500
Test	NA	NA	NA	6,500

Table 2: Distribution of labels over the training, validation and test sets for Task 6.

3 Our approach

For both tasks we implemented models based on the deep neural *transformer* architecture (Vaswani et al., 2017), since they have achieved state-of-the-art (SOTA) results in several NLP tasks. We studied the impact of fine-tuned BERT Large and CT-BERT pretrained models (Devlin et al., 2018)(Müller et al., 2020). For both models we employed the Pytorch implementation available from the HuggingFace library (Wolf et al., 2020).

3.1 Model Architecture

Figure 1 shows the general architecture shared by the two used models. It follows a standard design for sentence classification tasks using BERT, which considers the hidden state h of the final layer over the special token [CLS] as the full representation of the input sequences, and on top of this a classifier. The classifier head consists of a fully-connected layer with dropout probability of 0.1, 1024 input units, with 2 output units for Task 5 and 3 units for Task 6, followed by a softmax activation function to predict the class probability given the hidden state representation.

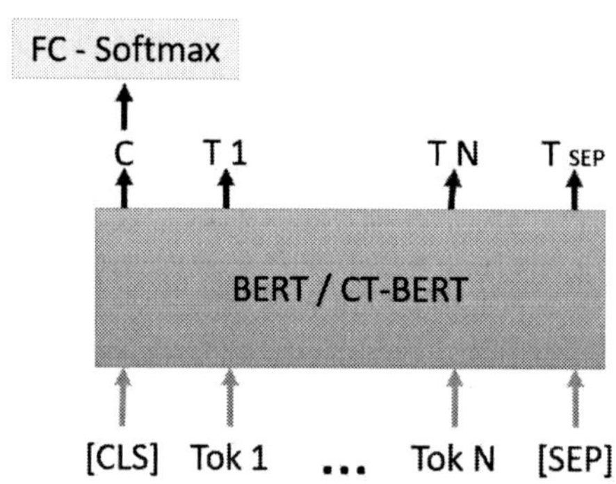

Figure 1: General model architecture

4 Experimental Setup

In this section we describe the training process for the two models in both tasks.

4.1 Data Preprocessing

All tweets in both tasks were preprocessed with the following operations[1]:

- Replace @usernames with a "user" token.
- Replace multiple occurrences of the "user" token with "n user", where n denotes the number of times the "user" token appears in the tweet.

[1] https://github.com/digitalepidemiologylab/covid-twitter-bert/tree/master/utils

- Replace URLs with a "url" token.
- Replace multiple occurrences of the "url" token with "n url", where n denotes the number of times the "url" token appears in the tweet.
- Convert emojis to their text aliases using the emoji library[2].
- Standardize text to ASCII representation. Using the Unidecode library[3], we removed all unicode symbols, punctuation and accented characters.
- Lowercase all the text.

4.2 Fine-tuning of models

In both tasks the models were fine-tuned with the Optuna framework (Akiba et al., 2019) using a random search approach by trying 10 different combinations for each model in each task. The search space described in Table 3 was defined so that the range of values stay close to the recommended values for BERT[4].

Weight Decay	LR	Epochs
0 - 0.3	1e-5 - 5e-5	1 - 4

Table 3: Hyper parameter search space

For Task 5, the best performing hyper parameter set according to the F1 score in the validation set was used in both models: 0.1328 *weight decay*, 3.154e-5 *learning rate* and 3 *epochs*. For Task 6 the values selected were 0.1423 *weight decay*, 3.278e-5 *learning rate* and 3 *epochs* for both models. Furthermore, the models for both tasks were trained over the concatenation of the training and validation sets for the final submission.

4.3 Results

Performance was measured using precision, recall and macro F1 metrics. Tables 4 and 5 show the performance attained in the test and validation sets of Task 5 and 6 respectively. Results in the validation and test sets of Task 5 were better for the CT-BERT model by a large margin, whilst for Task 6 both models achieved a high score with a small difference between them. CT-BERT outperformed both the BERT and the median submission score of all participants in the test set in both tasks achieving the highest score of all systems in Task 6.

[2]https://pypi.org/project/emoji/
[3]https://pypi.org/project/Unidecode/
[4]https://github.com/google-research/bert

System	Data	F1	P	R
BERT	Val	0.82	0.83	0.81
CT-BERT	Val	**0.89**	**0.89**	**0.90**
BERT	Test	0.68	0.71	0.65
CT-BERT	Test	**0.77**	**0.76**	**0.77**
Median	Test	0.74	0.73	0.74

Table 4: Results on test and validation data for Task 5

System	Data	F1	P	R
BERT	Val	**0.99**	**0.99**	**0.99**
CT-BERT	Val	0.98	0.98	0.98
BERT	Test	0.94	0.93	0.93
CT-BERT	Test	**0.95**	**0.94**	**0.94**
Median	Test	0.93	0.93	0.93

Table 5: Results on test and validation data for Task 6

5 Discussion

The training set for Task 5 was unbalanced, approximately at a ratio of 1:5 for classes "1" and "0". During experimentation an attempt was made to balance the classes to see if this would improve the results, but this approach was abandoned as the metrics dropped considerably. Also, it was observed that data preprocessing had a greater impact than hyper parameter search on the models for both tasks.

We analyzed the errors on the validation set in Task 5 for both models. The BERT model mislabeled 60 tweets vs 44 for CT-BERT, with 27 errors in common. Next we show two tweets that both models wrongly predicted:

- *i cough once and people think i have the coronavirus.* **Predicted = 1, True label = 0**
- *I legit feel super sick to my stomach and really weak hopefully I'm not dying from coronavirus.* **Predicted = 0, True label = 1**

Data in Task 5 showed that tweets labeled as "1" contain more mentions of the words "i", "got", "cough" and other symptoms compared to the tweets labeled as "0". While analyzing the common errors in both BERT and CT-BERT on the validation set, we discovered that "0" tweets that included these words were often misclassified as "1". On the other hand, in the validation set of Task 6 the BERT model mislabeled 4 tweets vs 9 tweets for CT-BERT, with only 3 errors in common. The following are two examples of errors made by our model.

- *@user Hi @user. The symptoms of Covid-19 are similar to that of a common cold or flu. These symptoms are: fatigue, fever, coughing, stuffy nose, sore throat or diarrhea. Seek medical attention if you, your child or family member show any of these signs. url.* **Predicted = Lit-News, True label = Nonpersonal Reports**

- *@user1 @user2 @user3 @user4 @user5 @user6 @user7 @user8 @user9 Man life is full of tortures. Has everyone with covid19 shown excessive damage in India? It is the opposite. 80% are asymptomatic. 10% have fever and 5% require medical supervision and rest need oxygen support. No need to panic..* **Predicted = Nonpersonal Reports, True label = Lit-News**

All errors in the validation set were misclassifications between the Lit-News and Nonpersonal Reports labels, where the model struggles to differentiate between the size of the audience of the tweet. In addition, tweets written in an impersonal style tend to be classified as News, whereas tweets written in first person tend to be classified as Nonpersonal or Self Reports.

6 Conclusions and future work

Transfer learning has shown to achieve above average results for various NLP tasks, where domain specific models can attain better results even if they were trained with less data than a general domain model. Based on the results obtained, we conclude that a model trained with quality domain-specific data (CT-BERT) can outperform a model trained with a much larger amount of general domain data (BERT).

In the experimental stage we also considered the BioBERT model, which was trained on a medical corpus (Lee et al., 2019), but it was not possible due to time constraints. Thus, we envision a potential future work to further compare the reach of several domain-specific models in the prediction of social media posts that could embed Covid-19 infection risk information.

Based on the error analysis, we plan to further improve our models' performance by considering *wide & deep* learning techniques (Cheng et al., 2016), which help to enhance their generalization ability by adding other types of features through the wide branch.

References

Takuya Akiba, Shotaro Sano, Toshihiko Yanase, Takeru Ohta, and Masanori Koyama. 2019. Optuna: A next-generation hyperparameter optimization framework.

Heng-Tze Cheng, Levent Koc, Jeremiah Harmsen, Tal Shaked, Tushar Chandra, Hrishi Aradhye, Glen Anderson, Greg Corrado, Wei Chai, Mustafa Ispir, Rohan Anil, Zakaria Haque, Lichan Hong, Vihan Jain, Xiaobing Liu, and Hemal Shah. 2016. Wide deep learning for recommender systems.

Jacob Devlin, Ming-Wei Chang, Kenton Lee, and Kristina Toutanova. 2018. Bert: Pre-training of deep bidirectional transformers for language understanding.

Ari Z Klein, Arjun Magge, Karen O'Connor, Jesus Ivan Flores Amaro, Davy Weissenbacher, and Graciela Gonzalez Hernandez. 2021. Toward using twitter for tracking covid-19: A natural language processing pipeline and exploratory data set. *J Med Internet Res*, 23(1):e25314.

Jinhyuk Lee, Wonjin Yoon, Sungdong Kim, Donghyeon Kim, Sunkyu Kim, Chan Ho So, and Jaewoo Kang. 2019. BioBERT: a pre-trained biomedical language representation model for biomedical text mining. *Bioinformatics*, 36(4):1234–1240.

Arjun Magge, Ari Klein, Ivan Flores, Ilseyar Alimova, Mohammed Ali Al-garadi, Antonio Miranda-Escalada, Zulfat Miftahutdinov, Eulàlia Farré-Maduell, Salvador Lima López, Juan M Banda, Karen O'Connor, Abeed Sarker, Elena Tutubalina, Martin Krallinger, Davy Weissenbacher, and Graciela Gonzalez-Hernandez. 2021. Overview of the sixth social media mining for health applications (# smm4h) shared tasks at naacl 2021. In *Proceedings of the Sixth Social Media Mining for Health Applications Workshop & Shared Task.*

Martin Müller, Marcel Salathé, and Per E Kummervold. 2020. Covid-twitter-bert: A natural language processing model to analyse covid-19 content on twitter.

Ashish Vaswani, Noam Shazeer, Niki Parmar, Jakob Uszkoreit, Llion Jones, Aidan N. Gomez, Lukasz Kaiser, and Illia Polosukhin. 2017. Attention is all you need.

Thomas Wolf, Lysandre Debut, Victor Sanh, Julien Chaumond, Clement Delangue, Anthony Moi, Pierric Cistac, Tim Rault, Rémi Louf, Morgan Funtowicz, Joe Davison, Sam Shleifer, Patrick von Platen, Clara Ma, Yacine Jernite, Julien Plu, Canwen Xu, Teven Le Scao, Sylvain Gugger, Mariama Drame, Quentin Lhoest, and Alexander M. Rush. 2020. Huggingface's transformers: State-of-the-art natural language processing.

System description for ProfNER - SMMH: Optimized fine tuning of a pretrained transformer and word vectors

David Fidalgo and **Daniel Vila-Suero** and **Francisco Aranda** and **Ignacio Talavera**
Recognai
https://www.recogn.ai
[david, daniel, francisco]@recogn.ai
ignaciotalaveracepeda@gmail.com

Abstract

This shared task system description depicts two neural network architectures submitted to the ProfNER track, among them the winning system that scored highest in the two sub-tasks 7a and 7b. We present in detail the approach, preprocessing steps and the architectures used to achieve the submitted results, and also provide a GitHub repository to reproduce the scores. The winning system is based on a transformer-based pretrained language model and solves the two sub-tasks simultaneously.

1 Introduction

The identification of professions and occupations in Spanish (ProfNER[1], Miranda-Escalada et al. 2021) is part of the Social Media Mining for Health Applications (SMM4H) Shared Task 2021 (Magge et al., 2021). Its aim was to extract professions from social media to enable characterizing health-related issues, in particular in the context of COVID-19 epidemiology as well as mental health conditions.

ProfNER was the seventh track of the task and focused on the identification of professions and occupations in Spanish tweets. It consisted of two sub-tasks:

- **task 7a:** In this binary classification task, participants had to determine whether a tweet contains a mention of occupation, or not.
- **task 7b:** In this Named Entity Recognition (NER) task, participants had to find the beginning and end of occupation mentions and classify them into two categories: `PROFESION` (professions) and `SITUACION_LABORAL` (working status).

2 Our approach

We submitted two systems to each of the tasks described above, which share the same basic structure:

- a *backbone* model that extracts and contextualizes the input features
- a *task head* that performs task specific operations and computes the loss

In the backbone of both systems we take advantage of pretrained components, such as a transformer-based language model or skip-gram word vectors. The task head of both systems is very similar in that it solves task 7a and 7b simultaneously, and returns the sum of both losses.

For the first system we aimed to maximize the metrics of the competition with the constraint of using a single GPU environment. For the second system we tried to maximize the model's efficiency with respect to the model size and speed while maintaining acceptable performance.

Both systems were designed and trained using *biome.text*[2], a practical NLP open source library based on AllenNLP[3] (Gardner et al., 2017) and PyTorch[4] (Paszke et al., 2019).

2.1 Preprocessing

In a first step we transformed the given *brat*[5] annotations of task 7b to commonly used *BIO* NER tags(Ratinov and Roth, 2009). For this we used spaCy[6] (Honnibal et al., 2020) and a customized tokenizer of its "`es_core_news_sm`" language model, to make sure that the resulting word tokens and annotations always aligned well. In this step we excluded the entity classes not considered during evaluation. The same customized tokenizer was used to transform the predicted NER tags of our systems back to brat annotations during inference time.

[1] `https://temu.bsc.es/smm4h-spanish/`

[2] `https://www.recogn.ai/biome-text`
[3] `https://allennlp.org/`
[4] `https://pytorch.org/`
[5] `http://brat.nlplab.org`
[6] `https://spacy.io/`

Proceedings of the Sixth Social Media Mining for Health Workshop 2021, pages 69–73
June 10, 2021.

tweet ID	word tokens	NER tags	classification label
1242604595463032832	[El, alcalde, ...]	[O, B-PROFESION, ...]	1
1242603450321506304	[", Trump, decide, ...]	[O, O, O, ...]	0
...	...	...	...

Table 1: Example of the format of our input data. NER tags are provided in the BIO encoding scheme.

To obtain the input data for our training pipeline, we added the tweet ID and the corresponding classification labels of task 7a to our word tokens and NER tags (see Table 1 for an example).

No data augmentation or external data was used for the training of our systems.

2.2 System 1: Transformer

In our first system, the backbone model consists of a transformer-based pretrained language model. More precisely, we use *BETO*, a BERT model trained on a big Spanish corpus (Cañete et al., 2020), which is distributed via Hugging Face's (Wolf et al., 2019) Model Hub[7] under the name "`dccuchile/bert-base-spanish-wwm-cased`". For its usage we further tokenize the word tokens into word pieces with the corresponding BERT tokenizer, which also introduces the special BERT tokens `[CLS]` and `[SEP]` (Devlin et al., 2019). Since some of the word tokens cannot be processed by the tokenizer and are simply ignored (e.g. the newline character "`\n`"), we replace those problematic word tokens with a dummy token "æ", which is not ignored, and that allows the correct transformation of NER tags to brat annotations at inference time. The output sequence of the transformer is then passed on to the task head of the system.

In the task head we first apply a non-linear tanh activation layer to the `[CLS]` token, which we initialize with its pretrained weights (Devlin et al., 2019), before obtaining the logits of a linear classification layer that solves task 7a. The classification loss is calculated via the Cross Entropy loss function. To solve task 7b, we need to bridge the difference between the word piece features and predictions at a the level of word tokens. For this, we follow the approach of Devlin et al. (2019) who use a subword pooling in which the first word piece of a word token is used to represent the entire token, excluding the special BERT tokens. After the subword pooling we apply a linear classification layer and a subsequent Conditional Random Field (CRF) model that predicts a sequence of NER tags.

parameter	search space
learning rate	loguniform(5e-6, 1e-4)
weight decay	loguniform(1e-3, 1e-1)
warm-up steps	randint(0, 200)
batch size	choice([8, 16])
num epochs	choice([3, 4, 5])

Table 2: List of hyperparameters tuned during training. Search spaces define valid values for the hyperparameters and how they are sampled initially. They are provided as Ray Tune search space functions.

2.2.1 Training

For the parameter updates we used the AdamW algorithm (Loshchilov and Hutter, 2019) and schedule the learning rate with warm-up steps and a linear decay afterwards. We optimized the training parameters listed in Table 2 by means of the *Ray Tune* library[8] (Liaw et al., 2018) which is tightly integrated with *biome.text*. Our Hyperparameter Optimization (HPO) consisted of 50 runs (see Figure 1) using a tree-structured Parzen Estimator[9] as search algorithm (Bergstra et al., 2011) and the ASHA trial scheduler to terminate low-performing trials (Li et al., 2018). The reference metric for both algorithms was the overall F1 score of task 7b. The HPO lasted for about 6 hours on a *g4dn.xlarge* AWS machine with one Tesla T4 GPU.

We took the best performing model of the HPO, performed a quick sweep across several random seeds for the initialization[10] and finally employed the best configuration to train the system on the combined train and validation data set.

In further experiments, we tried to improve the validation metrics by switching to BILOU tags (Ratinov and Roth, 2009) or by including the entity classes not considered for the final evaluation, but could not find any significance differences.

[7] `https://huggingface.co/models`

[8] `https://docs.ray.io/en/master/tune/`

[9] `https://github.com/hyperopt/hyperopt`

[10] In hindsight, it would have been better to perform this sweep before the HPO and include the best performing random seeds in the HPO itself.

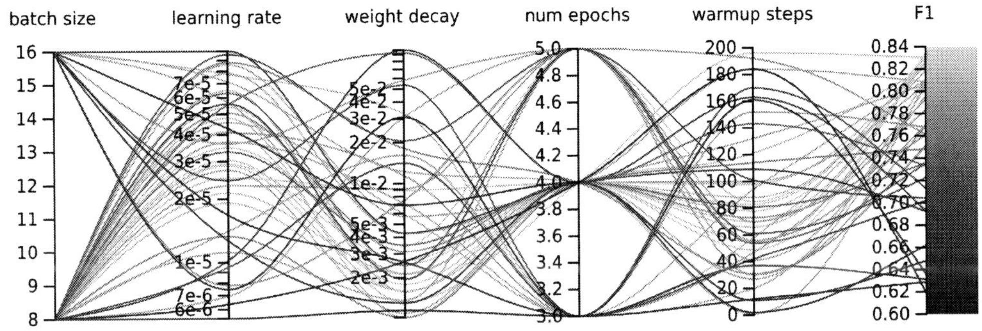

Figure 1: Distribution of the hyperparameters during the HPO for system 1. In total we executed 50 trials using a tree-structured Parzen Estimator as search algorithm and the ASHA trial scheduler to terminate low-performing trials early. The trial with the highest F1 NER score had a batch size of 8, a learning rate of 3.03e-05, a weight decay of 1.79e-3, was trained for 4 epochs and had 49 warm-up steps.

2.3 System 2: RNN

In our second system, the backbone model extracts word and character features, and combines them at a word token level. For the word feature we start from a cased version of skip-gram word vectors that were pretrained on 140 million Spanish tweets[11]. We concatenate these word vectors with the output of the last hidden state of a bidirectional Gated Recurrent Unit (GRU, Cho et al., 2014) that takes as input the lower cased characters of a word token. These embeddings are then fed into another larger bidirectional GRU, where we add contextual information to the encoding, and whose hidden states are passed on to the task head of the system.

In the task head we pool the sequence by means of a bidirectional Long short-term memory (LSTM, Hochreiter and Schmidhuber, 1997) unit and pass the last hidden state to a classification layer to solver task 7a. The classification loss is calculated via the Cross Entropy loss function. To solve task 7b, we pass each embedding from the backbone sequence through a feedforward network with a linear classification layer on top. The outputs of the classification layer are fed into a CRF model that predicts a sequence of NER tags.

The architectural choice of using GRU or LSTM units was solved via an HPO as described in the following training subsection.

2.3.1 Training

For the parameter updates we apply the same optimization algorithm and learning rate scheduler as for system 1. The comparatively small size of system 2 allowed us to perform extensive HPOs, not only for the training parameters but also for the architecture, and to some extent Neural Architecture Searches (NAS).

Backbone	
Word feature	300 dim, pretrained word vectors
Char feature	64 dim char vectors pooled by a GRU (bidirectional, 1 layer, 64 hidden size)
Backbone encoder	GRU (bidirectional, 1 layer, 512 hidden size)
Task Head	
Classification pooler	LSTM (bidirectional, 1 layer, 64 hidden size)
Feedforward	1 layer, 128 hidden size

Table 3: Details of our best RNN architecture.

In a first optimization run of 200 trials, we allowed wide ranges for almost all hyperparameters and tried out different RNN architectures, that is either LSTMs or GRUs. An example of a clearly preferred choice are the word embeddings pretrained with a skip-gram model over the ones pretrained with a a CBOW model (Mikolov et al., 2013). In a second run, we fixed obviously preferred choices and narrowed down the search spaces to the most promising ones.

For both HPO runs we applied the same search algorithm and trial scheduler as for system 1, and proceeded the same way to obtain the submitted version of system 2.

The resulting best RNN architecture is detailed in Table 3.

[11] https://zenodo.org/record/4449930

System	F1 Test (task 7a)	F1 Test (task 7b)	F1 Valid. (task 7a)	F1 Valid. (task 7b)	Model size (nr of params)	Inference time* (for 1 prediction)
1: Transformer	0.93	0.839	0.92	0.834	$\sim 1.1 \times 10^8$	24.5 ms $\pm$ 854 µs
2: RNN	0.88	0.764	0.85	0.731	$\sim 1.5 \times 10^7$	3.7 ms $\pm$ 103 µs

Table 4: Results for the two systems. Test results are provided with the systems trained on the combined training and validation data set, while the validation metric is taken from the best performing HPO trial. System 1 was the winning system in both ProfNER sub-tracks, while system 2 still scored above the arithmetic median of 0.85 and 0.7605 in both tasks.
*Mean value, computed on an i7-9750 H CPU with 6 cores.

3 Results

Table 4 presents the evaluation metrics of both systems on the validation and the test data sets, as well as the model size and its inference speed. With system 1 we managed to score highest on both ProfNER 7a and 7b sub-tracks (F1:0.93/P:0.9251/R:0.933 and F1:0.839/P:0.838/R:0.84, respectively), with an average of 8 points above the arithmetic median of all submissions. The much smaller and faster (by a factor of $\sim$ 7) system 2 still manages to score above the competitions median (F1:0.88/P:0.9083/R:0.8553 and F1:0.764/P:0.815/R:0.718, respectively), but performs significantly worse when compared to system 1.

We find a clear correlation between the classification F1 score and the F1 score of the NER task in our HPO runs, which signals that the feedback loop between the two tasks is in general beneficial and advocates solving both tasks simultaneously.

When comparing system 1 and 2, it seems that the amount of training data provided to the RNN architecture was not sufficient to match the transfer capabilities of the pretrained transformer, even with dedicated architecture searches and extensive hyperparameter tuning. This is corroborated by the fact that adding the validation data to the training data led to a clear performance boost for system 2, while the performance of system 1 stayed almost the same (compare the F1 Test and Validation metrics for task 7b in Table 4).

A possible path to improve system 1, which was not pursued due to time constraints, could be the inclusion of the gazetteers provided during the ProfNER track. We consider this path especially promising given the fact that the precision was always lower than the recall for both tasks.

We conclude that the exploitation of the transfer capabilities of a pretrained language model and its optimized fine tuning to the target domain, provides an conceptually easy system architecture and seems to be the most straight forward method to achieve competitive performance, especially for tasks where training data is scarce.

To help to reproduce our results, we provide a GitHub repository at `https://github.com/recognai/profner`.

Acknowledgments

This work was supported by the Spanish Ministerio de Ciencia, Inonvacion y Universidades through its Ayuda para contratos Torres Quevedo 2018 program with the reference number PTQ2018-009909.

References

James Bergstra, R. Bardenet, Yoshua Bengio, and Balázs Kégl. 2011. Algorithms for Hyper-Parameter Optimization. In *25th Annual Conference on Neural Information Processing Systems (NIPS 2011)*, volume 24 of *Advances in Neural Information Processing Systems*, Granada, Spain. Neural Information Processing Systems Foundation.

José Cañete, Gabriel Chaperon, Rodrigo Fuentes, Jou-Hui Ho, Hojin Kang, and Jorge Pérez. 2020. Spanish pre-trained bert model and evaluation data. In *PML4DC at ICLR 2020*.

Kyunghyun Cho, Bart van Merrienboer, Caglar Gulcehre, Dzmitry Bahdanau, Fethi Bougares, Holger Schwenk, and Yoshua Bengio. 2014. Learning phrase representations using rnn encoder-decoder for statistical machine translation.

Jacob Devlin, Ming-Wei Chang, Kenton Lee, and Kristina Toutanova. 2019. Bert: Pre-training of deep bidirectional transformers for language understanding.

Matt Gardner, Joel Grus, Mark Neumann, Oyvind Tafjord, Pradeep Dasigi, Nelson F. Liu, Matthew Peters, Michael Schmitz, and Luke S. Zettlemoyer. 2017. Allennlp: A deep semantic natural language processing platform.

S. Hochreiter and J. Schmidhuber. 1997. Long short-term memory. *Neural Computation*, 9:1735–1780.

Matthew Honnibal, Ines Montani, Sofie Van Landeghem, and Adriane Boyd. 2020. spaCy: Industrial-strength Natural Language Processing in Python.

Liam Li, Kevin Jamieson, Afshin Rostamizadeh, Ekaterina Gonina, Moritz Hardt, Benjamin Recht, and Ameet Talwalkar. 2018. A System for Massively Parallel Hyperparameter Tuning. *arXiv e-prints*, page arXiv:1810.05934.

Richard Liaw, Eric Liang, Robert Nishihara, Philipp Moritz, Joseph E. Gonzalez, and Ion Stoica. 2018. Tune: A Research Platform for Distributed Model Selection and Training. *arXiv e-prints*, page arXiv:1807.05118.

Ilya Loshchilov and Frank Hutter. 2019. Decoupled weight decay regularization.

Arjun Magge, Ari Klein, Ivan Flores, Ilseyar Alimova, Mohammed Ali Al-garadi, Antonio Miranda-Escalada, Zulfat Miftahutdinov, Eulàlia Farré-Maduell, Salvador Lima López, Juan M Banda, Karen O'Connor, Abeed Sarker, Elena Tutubalina, Martin Krallinger, Davy Weissenbacher, and Graciela Gonzalez-Hernandez. 2021. Overview of the sixth social media mining for health applications (# smm4h) shared tasks at naacl 2021. In *Proceedings of the Sixth Social Media Mining for Health Applications Workshop & Shared Task*.

Tomas Mikolov, Kai Chen, Greg Corrado, and Jeffrey Dean. 2013. Efficient estimation of word representations in vector space.

Antonio Miranda-Escalada, Eulàlia Farré-Maduell, Salvador Lima López, Luis Gascó-Sánchez, Vicent Briva-Iglesias, Marvin Agüero-Torales, and Martin Krallinger. 2021. The profner shared task on automatic recognition of occupation mentions in social media: systems, evaluation, guidelines, embeddings and corpora. In *Proceedings of the Sixth Social Media Mining for Health Applications Workshop & Shared Task*.

Adam Paszke, Sam Gross, Francisco Massa, Adam Lerer, James Bradbury, Gregory Chanan, Trevor Killeen, Zeming Lin, Natalia Gimelshein, Luca Antiga, Alban Desmaison, Andreas Kopf, Edward Yang, Zachary DeVito, Martin Raison, Alykhan Tejani, Sasank Chilamkurthy, Benoit Steiner, Lu Fang, Junjie Bai, and Soumith Chintala. 2019. Pytorch: An imperative style, high-performance deep learning library. In H. Wallach, H. Larochelle, A. Beygelzimer, F. d'Alché-Buc, E. Fox, and R. Garnett, editors, *Advances in Neural Information Processing Systems 32*, pages 8024–8035. Curran Associates, Inc.

Lev Ratinov and Dan Roth. 2009. Design Challenges and Misconceptions in Named Entity Recognition. In *Proc. of the Conference on Computational Natural Language Learning (CoNLL)*.

Thomas Wolf, Lysandre Debut, Victor Sanh, Julien Chaumond, Clement Delangue, Anthony Moi, Pierric Cistac, Tim Rault, Rémi Louf, Morgan Funtowicz, Joe Davison, Sam Shleifer, Patrick von Platen, Clara Ma, Yacine Jernite, Julien Plu, Canwen Xu, Teven Le Scao, Sylvain Gugger, Mariama Drame, Quentin Lhoest, and Alexander M. Rush. 2019. HuggingFace's Transformers: State-of-the-art Natural Language Processing. *arXiv e-prints*, page arXiv:1910.03771.

Word Embeddings, Cosine Similarity and Deep Learning for Identification of Professions & Occupations in Health-related Social Media

Sergio Santamaría Carrasco and **Roberto Cuervo Rosillo**
Universidad Carlos III de Madrid,
Computer Science Department,
Av de la Universidad, 30,
28911, Leganés, Madrid,Spain
sesantam@pa.uc3m.es rcuervo@pa.uc3m.es

Abstract

ProfNER-ST focuses on the recognition of professions and occupations from Twitter using Spanish data. Our participation is based on a combination of word-level embeddings, including pre-trained Spanish BERT, as well as cosine similarity computed over a subset of entities that serve as input for an encoder-decoder architecture with attention mechanism. Finally, our best score achieved an F1-measure of 0.823 in the official test set.

1 Introduction

During situations of risk, such as the Covid-19 pandemic, detecting vulnerable occupations, be it due to their risk of direct exposure to the threat or due to mental health issues associated with work-related aspects, is critical to prepare preventive measures. These occupations can be detected through the analysis of tweets since Twitter has become a very useful tool to find reliable information. Due to the exponential growth of the use of this social network, natural language processing (NLP) techniques have become a crucial tool for unlocking this critical information.

This paper describes the participation of our team in ProfNER-ST (Miranda-Escalada et al., 2021b) challenge, 7b subtrack of the sixth Social Media Mining for Health Applications (SMM4HA) (Magge et al., 2021), which focuses on the recognition of professions and occupations from Twitter using Spanish data.

The core of the proposed system is based on an encoder-decoder architecture with attention mechanism successfully applied previously (Ali and Tan, 2019) for temporal expression recognition. This system combines several neural network architectures for the extraction of characteristics at a contextual level and a CRF for the decoding of labels. The proposed system reach a F1 score of 0.823.

2 Methods and system description

2.1 Pre-processing

We pre-process the text of the clinical cases taking into account different steps. First, the corpus are clean from urls. Secondly, the tweets are split into tokens using Spacy[1], an open-source library that provides support for texts in several languages, including Spanish. Finally, the text and its annotations are transformed into the CoNLL-2003 format using the BIOES schema (Ratinov and Roth, 2009).

2.2 Features

- **Words**: Two different 300 dimensional representations based on pre-trained word embeddings has been used with FastText (Bojanowski et al., 2016). Both have been selected for their contribution of domain-specific knowledge since the former have been generated from Spanish medical corpora (Soares et al., 2020) and the latter have been trained with Spanish Twitter data related to COVID-19 (Miranda-Escalada et al., 2021a). Contextual embeddings generated with a fine-tuned BETO (Cañete et al., 2020) model are also included, as these word representations are dynamically informed by the surrounding words improving performance.

- **Part-of-speech**: This feature has been considered due to the significant amount of information it offers about the word and its neighbors. It can also help in word sense disambiguation. The PoS-Tagging model used was the one provided by the Spacy. An embedding representation of this feature is learned during training, resulting in a 40-dimensional vector.

- **Characters**: We also add character-level embeddings of the words, learned during train-

[1]https://spacy.io/

Proceedings of the Sixth Social Media Mining for Health Workshop 2021, pages 74–76
June 10, 2021. ©2021 Association for Computational Linguistics

ing and resulting in a 30-dimensional vector. These have proven to be useful for specific-domain tasks and morphologically-rich languages.

- **Syllables**: Syllable-level embeddings of the words, learned during training and resulting in a 75-dimensional vector is also added. Like character-level embeddings, they help to deal with words outside the vocabulary and contribute to capturing common prefixes and suffixes in the domain and correctly classifying words.
- **Cosine Similarity**: The BETO embeddings of the entities found in the training and validation set are used to calculate the cosine similarity between the BETO representation of the word to be analyzed, since previous work (Büyüktopaç and Acarman, 2019) has shown that could help to improve the results on data extracted from Twitter. This information is encoded as a 3717-dimensional vector.

2.3 Architecture

In the proposed system, shown in Figure 1, the character and syllable information is previously processed by a convolutional and global max pooling block, to be concatenated with the rest of the input features to serve as input to an encoder-decoder architecture with attention mechanism. The context vector as well as decoder outputs feeds a fully connected dense layer with $tanh$ activation function. The last layer (CRF optimization layer) consists of a conditional random fields layer selected due to the ability of the layer to take into account the dependencies between the different labels. The output of this layer provides the most probable sequence of labels.

The system has been developed in python 3 (Van Rossum and Drake, 2009) with Keras 2.2.4 (Chollet et al., 2015) and Tensorflow 1.14.0 (Abadi et al., 2016).

3 Results

During experimentation our team apply the standard measures, precision, recall, and micro-averaged F1-score, to evaluate the performance of our model.

While the training set (Miranda-Escalada et al., 2020) was used for training the model, the development set was exploited to hyperparameter fine tuning. In the prediction stage, we combined both sets

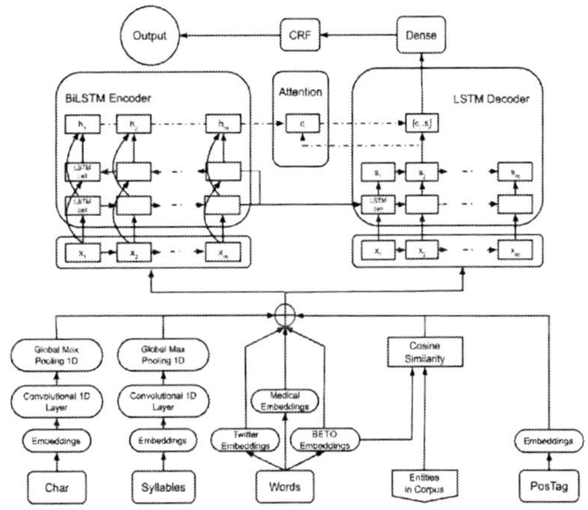

Figure 1: Architecture of the proposed model for profession and occupations recognition.

to training the model. The detailed hyper-parameter settings are illustrated in Table 1 'Opt.' denotes optimal.

Parameters	Tuned range	Opt
Train batch size	[8, 32, 64]	32
Epoch number	[2,3,4,5,6]	4
Dropout	[0.4, 0.5]	0.4
Max Seq Length	[50, 75, 100]	75
Learning rate	[0.01, 0.001, 0.0001]	0.001
Optimizer	-	Adam

Table 1: Hyper-parameters details.

With the optimal parametric configuration obtained during the experimentation, the model obtains the results shown in the Table 2.

	Precision	Recall	F-1 Score
Validation	0.893	0.753	0.817
Test	0.883	0.77	0.823

Table 2: Final results obtained in the competition.

4 Conclusion

In these working notes we describe our proposed system based on an encoder-decoder architecture with an attention mechanism powered by a combination of word embeddings that include pre-trained fine-tuned Spanish BERT embeddings.

Future work would explore different Data Augmentation techniques as well as other entities information, as companies or organizations, which could contain important information related to occupations.

References

Martín Abadi, Paul Barham, Jianmin Chen, Zhifeng Chen, Andy Davis, Jeffrey Dean, Matthieu Devin, Sanjay Ghemawat, Geoffrey Irving, Michael Isard, et al. 2016. Tensorflow: A system for large-scale machine learning. In *12th {USENIX} Symposium on Operating Systems Design and Implementation ({OSDI} 16)*, pages 265–283.

Mohammed NA Ali and Guanzheng Tan. 2019. Bidirectional encoder–decoder model for arabic named entity recognition. *Arabian Journal for Science and Engineering*, 44(11):9693–9701.

Piotr Bojanowski, Edouard Grave, Armand Joulin, and Tomas Mikolov. 2016. Enriching word vectors with subword information. *arXiv preprint arXiv:1607.04606*.

Onur Büyüktopaç and Tankut Acarman. 2019. Evaluation of cosine similarity feature for named entity recognition on tweets. In *International Conference on Man–Machine Interactions*, pages 125–135. Springer.

José Cañete, Gabriel Chaperon, Rodrigo Fuentes, Jou-Hui Ho, Hojin Kang, and Jorge Pérez. 2020. Spanish pre-trained bert model and evaluation data. In *PML4DC at ICLR 2020*.

Francois Chollet et al. 2015. Keras.

Arjun Magge, Ari Klein, Ivan Flores, Ilseyar Alimova, Mohammed Ali Al-garadi, Antonio Miranda-Escalada, Zulfat Miftahutdinov, Eulàlia Farré-Maduell, Salvador Lima López, Juan M Banda, Karen O'Connor, Abeed Sarker, Elena Tutubalina, Martin Krallinger, Davy Weissenbacher, and Graciela Gonzalez-Hernandez. 2021. Overview of the sixth social media mining for health applications (# smm4h) shared tasks at naacl 2021. In *Proceedings of the Sixth Social Media Mining for Health Applications Workshop & Shared Task*.

Antonio Miranda-Escalada, Marvin Aguero, and Martin Krallinger. 2021a. Spanish covid-19 twitter embeddings in fasttext. Funded by the Plan de Impulso de las Tecnologías del Lenguaje (Plan TL).

Antonio Miranda-Escalada, Vicent Briva-Iglesias, Eulàlia Farré, Salvador Lima López, Marvin Aguero, and Martin Krallinger. 2020. ProfNER corpus: gold standard annotations for profession detection in Spanish COVID-19 tweets. Funded by the Plan de Impulso de las Tecnologías del Lenguaje (Plan TL).

Antonio Miranda-Escalada, Eulàlia Farré-Maduell, Salvador Lima López, Vicent Briva-Iglesias, Marvin Agüero-Torales, Luis Gascó-Sánchez, and Martin Krallinger. 2021b. The profner shared task on automatic recognition of professions and occupation mentions in social media: systems, evaluation, guidelines, embeddings and corpora. In *Proceedings of the Sixth Social Media Mining for Health Applications Workshop & Shared Task*.

Lev Ratinov and Dan Roth. 2009. Design challenges and misconceptions in named entity recognition. In *Proceedings of the Thirteenth Conference on Computational Natural Language Learning (CoNLL-2009)*, pages 147–155.

Felipe Soares, Marta Villegas, Aitor Gonzalez-Agirre, Jordi Armengol-Estapé, Siamak Barzegar, and Martin Krallinger. 2020. Fasttext spanish medical embeddings. Funded by the Plan de Impulso de las Tecnologías del Lenguaje (Plan TL).

Guido Van Rossum and Fred L. Drake. 2009. *Python 3 Reference Manual*. CreateSpace, Scotts Valley, CA.

A Online Resources

The sources for the Troy&AbedInTheMorning participation are available via

- GitHub `https://github.com/ssantamaria94/ProfNER-SMM4H`,

Classification, Extraction, and Normalization : CASIA_Unisound Team at the Social Media Mining for Health 2021 Shared Tasks

Tong Zhou[1,3,*,†], Zhucong Li[1,2,*], Zhen Gan[1,4,*,†], Baoli Zhang[1], Yubo Chen[1,2], Kun Niu[3]
Jing Wan[4], Kang Liu[1,2], Jun Zhao[1,2], Yafei Shi[5], Weifeng Chong[5], Shengping Liu[5]

[1] National Laboratory of Pattern Recognition, Institute of Automation, Chinese Academy of Sciences
[2] School of Artificial Intelligence University of Chinese Academy of Sciences
[3] Beijing University of Posts and Telecommunications
[4] Beijing University of Chemical Technology
[5] Beijing Unisound Information Technology Co., Ltd

{tongzhou21, niukun}@bupt.edu.cn
{zhucong.li,baoli.zhang,yubo.chen,kliu,jzhao}@nlpr.ia.ac.cn
{ganzhen, wanj}@mail.buct.edu.cn
{shiyafei,chongweifeng,liushengping}@unisound.com

Abstract

This is the system description of the CASIA_Unisound team for Task 1, Task 7b, and Task 8 of the sixth Social Media Mining for Health Applications (SMM4H) shared task in 2021. To address two shared challenges among those tasks, the colloquial text and the imbalance annotation, we apply customized pre-trained language models and propose various training strategies. Experimental results show the effectiveness of our system. Moreover, we got an F1-score of 0.87 in task 8, which is the highest among all participates.

1 Introduction

Enormous data in social media has drawn much attention in medical applications. With the rapid development of health language processing, effective systems in mining health information from social media were built to assist pharmacy, diagnosis, nursing, and so on (Paul et al., 2016) (Yang et al., 2012) (Zhou et al., 2018).

The health language processing lab at the University of Pennsylvania organized the Social Media Mining for Health Applications (SMM4H) shared task 2021 (mag), which provided an opportunity for fair competition among state-of-the-art health information mining systems customized in the social media domain. We participated in task 1, subtask b of task 7, and task 8.

Task 1 consists of three subtasks in a cascade manner: (1) identifying whether a tweet mentions adverse drug effect; (2) mark the exact position that mentions ADE in the tweet; (3) normalization ADE mentions to standard terms. Subtask b of task 7 (Miranda-Escalada et al., 2021) is designed to identify professions and occupations (ProfNER) in Spanish tweets during the COVID-19 outbreak. Task 8 is targeting the classification of self-reported breast cancer posts on Twitter.

The ubiquitous two challenges of all the SMM4H shared tasks are (1) how to properly model the colloquial text in tweets; (2) avoid prediction bias caused by learning from unbalanced annotated data. The tweet's text, mixing with informal spelling, various emojis, usernames mentioned, and hyperlinks, will hinder the real semantic comprehension by a common pre-trained language model. Meanwhile, medical concepts are imbalanced in the real world due to the imbalanced morbidity of various diseases, and this phenomenon is also reflected in social media data. Training with imbalanced data will induce the model to pay much attention to the major classes and neglect the tail classes, which hinders the model's robustness and generalization.

To address the challenges above, we utilize a language model pre-trained on tweet data as the backbone and introduce multiple data construction methods in the training process. In the following, we will describe our methods and corresponding experiments for each task separately. At last, we summary this competition and discuss future directions.

2 Task 1: English ADE Tweets Mining

Adverse drug effect (ADE) is among the leading cause of morbidity and mortality. The collection of those adverse effects is crucial in prescribing and new drug research.

* Equal contribution.

† Works are done during internship at National Laboratory of Pattern Recognition, Institute of Automation, Chinese Academy of Sciences.

Proceedings of the Sixth Social Media Mining for Health Workshop 2021, pages 77–82
June 10, 2021.

Tweet	MedDRA Term
vyvanse completely gets rid of my appetite. not quite sure how to feel about this.	10003028 appetite lost

Table 1: An example of tweets labeled ADE in Task 1. The ADE span is colored red, and the corresponding MedDRA term id is 10003028.

This task's objective is to find the tweet containing ADE, locate the span, and finally map the span to concepts in standard terms.

2.1 Classification

The goal of this subtask is to distinguish whether a tweet mentions adverse drug effects. As shown in Table 1, "rid of my appetite" is an ADE mention, so this tweet is labeled on "ADE". In this dataset, the training set consists of 17385 tweets (16150 NoADE and 1235 ADE tweets), the validation set consists of 914 labeled tweets (849 NoADE and 65 ADE tweets), and the test set consists of 10984 tweets. Since only about 7% of the tweets contain ADEs, we target this class imbalance issue with a customized pseudo data construction strategy.

2.1.1 Method

Pseudo Data: A human may differentiate ADE tweets by some complaints trigger words like verb "feel" "think" or some negative sentiment words like "gets rid of", but a more precise way is discerning ADE mention. The mention in the tweet indicating ADE is a colloquial MedDRA term, and they express the same semantic. We construct ADE tweet for training in two ways: (1) randomly inserting the text description of a standard term in a tweet; (2) regarding the text description of a standard term as an ADE tweet. With those pseudo training data, a model should pay more attention to ADE mention in a tweet and more robust to diversified and unseen context.

Model: We apply the BERTweet (Nguyen et al., 2020), a RoBERTa (Liu et al., 2019) language model pre-trained on Twitter data, to encode tweet text and make a binary prediction according to the corresponding pooling vector.

2.1.2 Experiments

We set the batch size to 32 and using AdamW (Loshchilov and Hutter, 2018) optimizer for optimizing. For BERTweet parameters, we set a learning rate of 3e-5, the weight of L2 normalization is 0.01; for other parameters, we set the learning rate to 3e-4, the weight of L2 normalization is 0. We finetune all models using 5-fold cross-validation on the training set for 50 epochs. The amount of pseudo data is equal to 85.80% of the origin training data to balance the two classes. The experimental results are shown in Table 2, and indicate the advantage of our data construction strategies.

Model	Precision	Recal	F1
Ours	0.592	0.417	0.49
Ours w/o pseudo data	0.552	0.325	0.41
Average scores	0.505	0.409	0.44

Table 2: Results on the SMM4H Task 1a test set.

2.2 Extraction

This subtask aims to extract ADE entities from English Twitter texts containing ADE. The dataset includes training set, validation set, and test set containing 17385, 915, and 10984 tweets respectively. The proportion of tweets involving ADE mentions in the training set and the validation set is about 7.1%.

2.2.1 Method

Preprocessing: To reflect real semantic properly, we preprocess tweets in customized manners. (1) Since most user names are outside the vocabulary, We change all user names behind @ to "user". (2) There are some escape characters in the Twitter text, such as """, "&", "<", ">", and we replace them with the original characters: """, "&", "<", ">" respectively.

Training: During the training stage, We use a five-fold cross-training fusion system, which include 7 different pre-training models. We ensemble them through average weighted voting to weaken the fluctuations of performance of single model.

Model: We use seven pre-training models: bertweet-base, bertweet-covid19-base-cased, bertweet-covid19-base-uncased, bert-base-cased, bert-base-uncased, bert-large-cased, and bert-large-uncased.

2.2.2 Experiments

The models we choose and their learning rates are shown in Table 3. Each model has two learning rates, the former is the learning rate of BERT, and the latter is the learning rate of BiLSTM(Ma and Hovy, 2016)+CRF(Lafferty et al., 2001). Each BERT model is finetuned for 50 epochs with the dropout (Srivastava et al., 2014) of 0.3 using AdamW (Loshchilov and Hutter, 2018) optimizer.

Model	Learning Rate
bertweet-base+BiLSTM+CRF	[5e-5, 5e-3]
bertweet-covid19-base-cased+ BiLSTM+CRF	[5e-5, 5e-3]
bertweet-covid19-base-uncased+ BiLSTM+CRF	[5e-5, 5e-3]
bert-base-cased+BiLSTM+CRF	[5e-5, 5e-3]
bert-base-uncased+BiLSTM+CRF	[4e-5, 4e-3]
bert-large-cased+CRF	[1e-5, 1e-3]
bert-large-uncased+CRF	[7e-6, 7e-4]

Table 3: Implementation details of our models of the SMM4H Task 1b.

Model	Precision	Recal	F1
Ours	0.381	0.475	0.42
Average scores	0.493	0.458	0.42

Table 4: Results on the SMM4H Task 1b test set.

We set the batch size of bert-large-cased and bert-large-uncased to 8, and the others are 64. The experimental results are shown in Table 4. The Recall of our result is close to two percentage points higher than the average, but our Precision is about 11 percentage points lower than the average. Therefore, our model recalls more correct entities, but it also recalls a lot of wrong entities. So this may be a direction in which our method can be optimized.

2.3 Normalization

MedDRA (Brown et al., 1999) is a rich and highly specific standardized medical terminology to facilitate sharing regulatory information internationally for medical products used by humans. This subtask aims to normalize ADE mention to standard MedDRA term based on the result of span detection.

2.3.1 Method

Our model's inference process consists of a classification phase and a compare phase, responsible for recall and rank, respectively. We train the above two phrases with shared parameters and optimizing with the combined supervising signal.

Recall: In view of the representation process of ADE's mention could be benefited from its context, we utilize BERTweet for complete tweet representation. Since we have a specific position of mention in a tweet from subtask b, we first truncate mention's representations and calculate out the mean vector as the mention representation. Next, we calculate the dot product between mention representation and term embedding. Each vector in the term embedding is initialized according to its corresponding mean BERTweet representation of standard term text description. Finally, a softmax operation is added to convert the dot product value to conditional probabilities. A cross-entropy loss function responsible for supervising this process.

Rank: Since the MedDRA term's description is a normalized expression of its corresponding ADE mention, the global semantic of a tweet should remain unchanged after exchanging the colloquial ADE mention and correct term description. On the contrary, the global semantic should have an offset after exchanging with a wrong term. Based on the above assumption, we add an additional supervising signal. A tweet's global representation is obtained from BERTweet's mean pooling vector. The model calculates triplet loss among the following global representations: (a) origin tweet (b) replace the mention with target term's description (c) replace the mention with a wrong term's description. The wrong term is firstly obtained by random selection from the whole term set, and with the procedures of the training process, it is randomly selected from the classification model's top K prediction. The triplet loss intends to maximize the similarity of the global representation of (a) and (b); meanwhile, it minimizes the similarity of (a) and (c).

Inference: In the inference stage, first, we obtain the top K terms based on the prediction of the recall procedure. Then we exchange the candidate K terms with the mention in the origin tweet and calculate the similarity of global representation with the origin tweet. The similarity score is the base of term ranking. Finally, we retain the top 1 as the final prediction.

Model	Precision	Recal	F1
Ours* (recall)	0.244	0.305	0.271
Ours* (recall + rank)	0.248	0.311	0.276
Ours (recall + rank)	0.129	0.403	0.195
Average scores	0.231	0.218	0.22

Table 5: Results on the SMM4H Task 1c test set, * denotes the results of our method based on our best prediction in subtask b.

2.3.2 Experiments

Our hyperparameter setting is identical to subtask a. Besides, we set K to 10, and for the combination of cross-entropy loss and triplet loss, we set equal weights. The experimental results are shown in Table 5, and indicate the advantage of the compare-based rank procedure.

3 Task 7: ProfNER for Spanish Tweets

3.1 Extraction

This subtask aims to detect the spans of professions and occupations entities in each Spanish tweet. The corpus contains four categories, but participants will only be evaluated to predict two of them: PROFESSION [profession] and SITUACION_LABORAL [working status]. The dataset includes a training set, validation set, and test set containing 6000, 2000, and 27000 tweets, respectively.

3.1.1 Method

Preprocessing: According to the characteristics of the competition's Spanish Twitter data and the competition requirements, we preprocess data to improve the model's ability to capture text information. (1) Since most user names are outside the vocabulary, We change all user names behind @ to "usuario". (2) The corpus contains four kinds of labels, but we will only be evaluated in the prediction of 2 of them: PROFESSION and SITUACION_LABORAL, so we removed the other two labels ACTIVIDAD and FIGURATIVA.

Training: Similar to subtask b of task 1, we make predictions on the multiple trained models and perform a simple voting scheme to get the final result.

Model: We use three BERT-based (Devlin et al., 2018) pre-training models: bert-base-spanish-wwm-cased, bert-spanish-cased-finetuned-ner, and bert-spanish-cased-finetuned-pos.

3.1.2 Experiments

For this subtask, each BERT model is finetuned for 50 epochs with the learning rate of 5e-5 using AdamW optimizer, and for the BiLSTM+CRF module, our learning rate is 5e-3, and the batch size is 64. The experimental results are shown in Table 6. The Model_ensemble0(noLSTM) is the result of the fusion of fifteen models without the BiLSTM modules, and The Model_ensemble1(LSTM) is the result of the fusion of fifteen models with the BiLSTM modules. The Ours is the final result, which is the voting fusion result of 30 models. From the experimental results, we can see that the F1 score of the fusion record on the validation set is superior, but the test set score has dropped. According to our analysis, this is probably related to a large amount of test data.

https://huggingface.co/dccuchile/bert-base-spanish-wwm-cased

https://huggingface.co/mrm8488

Model	Validation F1	Test F1
bert_spanish_cased	0.732	-
bert_spanish_ner	0.736	-
bert_spanish_pos	0.723	-
Model_ensemble0(noLSTM)	0.742	0.725
Model_ensemble1(LSTM)	0.744	0.731
Ours	-	0.733

Table 6: Results on the SMM4H Task 7b Validation and test set.

Tweet	**Label**
Excellent cause! I hope you are doing well. I had breast cancer too. I'm into my 3rd year of Tamoxifen.	S
OH MY GOD i just remembered my dream from my nap earlier i understand now why i felt so bad when i woke up i literally dreamt that i had breast cancer	NR

Table 7: Two examples of tweets and corresponding labels in Task 8.

4 Task 8: Self-reported Patient Detection

The adverse patient-centered outcomes (PCOs) caused by hormone therapy would lead to breast cancer patients discontinuing their long-term treatments (Fayanju et al., 2016). The research on PCOs is beneficial to reducing the risk of cancer recurrence. However, PCOs are not detectable through laboratory tests and are sparsely documented in electronic health records. Social media is a promising resource, and we can extract PCOs from the tweet with breast cancer self-reporting (Freedman et al., 2016). First and foremost, the PCO extraction system requires the accurate detection of self-reported breast cancer patients. This task's objective is to identify tweets in the self-reports category. In this dataset, the training set consists of 3513 tweets (898 self-report and 2615 non-relevant tweets), the validation set consists of 302 tweets (77 self-report and 225 non-relevant tweets), and the test set consists of 1204 tweets.

4.1 Method

Preprocessing: We preprocess the data to fit the tokenizer of the pre-trained RoBERTa model BERTweet, which is customized in tweet data. (1) The BERTweet's tokenizer transform the URL string in tweet to a unified special token by matching "http" or "www". For the tokenizer to effectively identify the URL, we insert "http://" before

Model	Precision	Recal	F1
Ours w/o preprocessing, w/o robust training	0.8571	0.8571	0.8571
Ours w/o robust training	0.8844	0.8442	0.8637
Ours	0.8701	0.8701	0.8701
Average scores	0.8701	0.8377	0.85

Table 8: Results on the SMM4H Task 8 test set.

"pic.twitter.com" in tweets. (2) The emoji in tweets is expressed as UTF-8 bytes code in string form. We match the "\x" and transform the code into its corresponding emoji.

Training: Although the generalization ability of the pre-trained language model finetuned in text classification tasks has been proved, it could still seize the wrong correction between specific tokens and the target label, turn out to neglect the crucial semantic. As shown at the top of Table 7, "I had breast cancer" is convincing evidence to a positive prediction. A model can make the right decision on the example at the bottom of Table 7 only if it takes the context into consideration. To avoid this wrong correction and improve our model's robustness, we apply two strategies on the training stage exert in data level and model level, respectively.

(1) Noise: Each word in a tweet has a probability p to be replaced by a random word, and the target label has a probability p to reverse.

(2) FGM: Following the fast gradient method (Miyato et al., 2016), we move the input one step further in the direction of rising loss, which will make the model loss rise in the fastest direction, thus forming an attack. In response, the model needs to find more robust parameters in the optimization process to deal with attacks against samples.

Model: Similar to subtask a in Task 1, we apply the BERTweet to encode tweet text and make a binary prediction according to the corresponding pooling vector.

4.2 Experiments

We set the batch size to 32 and using AdamW optimizer for optimizing. For BERTweet parameters, we set a learning rate of 3e-5, the weight of L2 normalization is 0.01; for other parameters, we set the learning rate to 3e-4, the weight of L2 normalization is 0. We set the noise rate to 0.025 and the epsilon of FGM to 0.5. We finetune all models using 5-fold cross-validation on the training set for 50 epochs. The experimental results are shown in Table 8. Our method has obtained the highest F1 score in this task. Furthermore, the ablation results indicate the advantage of the customized data preprocessing procedure and the robust training strategies.

5 Conclusion and Future Work

This work explores various customized methods in tasks of classification, extraction, and normalization of health information from social media. We have empirically evaluated different variants of our system and demonstrated the effectiveness of the proposed methods. As future work, we intend to introduce the medical domain's knowledge graph to improve our system further.

Acknowledgements

This work is supported by the National Key RD Program of China (2020AAA0106400), the National Natural Science Foundation of China (No.61806201) and the Key Research Program of the Chinese Academy of Sciences (Grant NO. ZDBS-SSW-JSC006).

References

Elliot G Brown, Louise Wood, and Sue Wood. 1999. The medical dictionary for regulatory activities (meddra). *Drug safety*, 20(2):109–117.

Jacob Devlin, Ming-Wei Chang, Kenton Lee, and Kristina Toutanova. 2018. Bert: Pre-training of deep bidirectional transformers for language understanding. *arXiv preprint arXiv:1810.04805*.

Oluwadamilola M Fayanju, Tinisha L Mayo, Tracy E Spinks, Seohyun Lee, Carlos H Barcenas, Benjamin D Smith, Sharon H Giordano, Rosa F Hwang, Richard A Ehlers, Jesse C Selber, et al. 2016. Value-based breast cancer care: a multidisciplinary approach for defining patient-centered outcomes. *Annals of surgical oncology*, 23(8):2385–2390.

Rachel A Freedman, Kasisomayajula Viswanath, Ines Vaz-Luis, and Nancy L Keating. 2016. Learning from social media: utilizing advanced data extraction techniques to understand barriers to breast cancer treatment. *Breast cancer research and treatment*, 158(2):395–405.

John D Lafferty, Andrew McCallum, and Fernando CN Pereira. 2001. Conditional random fields: Probabilistic models for segmenting and labeling sequence data. In *Proceedings of the Eighteenth International Conference on Machine Learning*, pages 282–289.

Yinhan Liu, Myle Ott, Naman Goyal, Jingfei Du, Mandar Joshi, Danqi Chen, Omer Levy, Mike Lewis, Luke Zettlemoyer, and Veselin Stoyanov. 2019. Roberta: A robustly optimized bert pretraining approach. *arXiv preprint arXiv:1907.11692*.

Ilya Loshchilov and Frank Hutter. 2018. Fixing weight decay regularization in adam.

Xuezhe Ma and Eduard Hovy. 2016. End-to-end sequence labeling via bi-directional lstm-cnns-crf. In *Proceedings of the 54th Annual Meeting of the Association for Computational Linguistics (Volume 1: Long Papers)*, pages 1064–1074.

Antonio Miranda-Escalada, Eulàlia Farré-Maduell, Salvador Lima López, Luis Gascó-Sánchez, Vicent Briva-Iglesias, Marvin Agüero-Torales, and Martin Krallinger. 2021. The profner shared task on automatic recognition of occupation mentions in social media: systems, evaluation, guidelines, embeddings and corpora. In *Proceedings of the Sixth Social Media Mining for Health Applications Workshop & Shared Task*.

Takeru Miyato, Andrew M Dai, and Ian Goodfellow. 2016. Adversarial training methods for semi-supervised text classification. *arXiv preprint arXiv:1605.07725*.

Dat Quoc Nguyen, Thanh Vu, and Anh Tuan Nguyen. 2020. Bertweet: A pre-trained language model for english tweets. *arXiv preprint arXiv:2005.10200*.

Michael J Paul, Abeed Sarker, John S Brownstein, Azadeh Nikfarjam, Matthew Scotch, Karen L Smith, and Graciela Gonzalez. 2016. Social media mining for public health monitoring and surveillance. In *Biocomputing 2016: Proceedings of the Pacific symposium*, pages 468–479. World Scientific.

Nitish Srivastava, Geoffrey Hinton, Alex Krizhevsky, Ilya Sutskever, and Ruslan Salakhutdinov. 2014. Dropout: a simple way to prevent neural networks from overfitting. *The journal of machine learning research*, 15(1):1929–1958.

Christopher C Yang, Haodong Yang, Ling Jiang, and Mi Zhang. 2012. Social media mining for drug safety signal detection. In *Proceedings of the 2012 international workshop on Smart health and wellbeing*, pages 33–40.

Lina Zhou, Dongsong Zhang, Christopher C Yang, and Yu Wang. 2018. Harnessing social media for health information management. *Electronic commerce research and applications*, 27:139–151.

Neural Text Classification and Stacked Heterogeneous Embeddings for Named Entity Recognition in SMM4H 2021

Usama Yaseen[1,2], Stefan Langer[1,2]
[1]Technology, Siemens AG Munich, Germany
[2]CIS, University of Munich (LMU) Munich, Germany
{usama.yaseen,langer.stefan}@siemens.com

Abstract

This paper presents our findings from participating in the SMM4H Shared Task 2021. We addressed Named Entity Recognition (NER) and Text Classification. To address NER we explored BiLSTM-CRF with Stacked Heterogeneous Embeddings and linguistic features. We investigated various machine learning algorithms (logistic regression, Support Vector Machine (SVM) and Neural Networks) to address text classification. Our proposed approaches can be generalized to different languages and we have shown its effectiveness for English and Spanish. Our text classification submissions (team:MIC-NLP) have achieved competitive performance with F1-score of 0.46 and 0.90 on ADE Classification (Task 1a) and Profession Classification (Task 7a) respectively. In the case of NER, our submissions scored F1-score of 0.50 and 0.82 on ADE Span Detection (Task 1b) and Profession Span detection (Task 7b) respectively.

1 Introduction

The ubiquity of social media has led to massive user-generated content across various platforms. Twitter is a popular micro-blogging platform that allows its users to publish tweets up to 280 characters. The common public uses Twitter to share life-related personal and professional experiences with others. Personal experiences often involve health-related incidents including mentions of adverse drug effect (ADE); this information is crucial to study Pharmacovigilance. In the context of the COVID-19 pandemic, the professional experiences may include information about professions and occupations which are vulnerable due to either direct exposure to the virus or due to the associated mental health issues; detecting vulnerable occupations is critical to adopt necessary preventive measures.

Recent research focuses on mining Twitter data for adverse drug effect detection (Jiang and Zheng, 2013; Adrover et al., 2015; Onishi et al., 2018). The distinctive style of communication on Twitter presents unique challenges including informal (brief) text, misspellings, noisy text, abbreviations, data sparsity, colloquial expressions and multilinguality.

2 Task Description and Contribution

We participate in the following two tasks organized by SMM4H workshop 2021 (Magge et al., 2021): (1) Task 1: Classification, Extraction and Normalization of Adverse Effect mentions in English tweets (2) Task 7: Identification of professions and occupations in Spanish tweets (Miranda-Escalada et al., 2021). Task 1 consists of three sub-tasks, (a): ADE tweet classification, (b): ADE span detection, (c): ADE resolution; whereas Task 7 consists of two sub-tasks: (a): Tweet classification (b): Profession/occupation span detection. For both tasks, we participate in sub-tasks (a) and (b). The Task 1a and Task 7a is a text classification problem while Task 1b and Task 7b is a Named Entity Recognition problem.

Following are our multi-fold contributions:

1. To address NER tasks, we have employed a neural network based sequence classifier, i.e. BiLSTM-CRF and investigated various heterogeneous embeddings. We further investigated the combination of character embeddings, static word embeddings and contextualized embeddings in a stacked format. We also incorporated linguistic features such as part-of-speech tags (POS), orthographic features etc. We apply the proposed modelling approaches to both English and Spanish texts. In *Profession span detection* (Task 7b) our submission (team:MIC-NLP) achieved the F1-score of 0.824 which is 6 points higher than the arithmetic median of all the submissions; in case of *ADE span detection* our submission scored F1-score of 0.50, around 8 points higher than the arithmetic median of the participating submissions.

2. To address text classification tasks, we investi-

Proceedings of the Sixth Social Media Mining for Health Workshop 2021, pages 83–87
June 10, 2021.

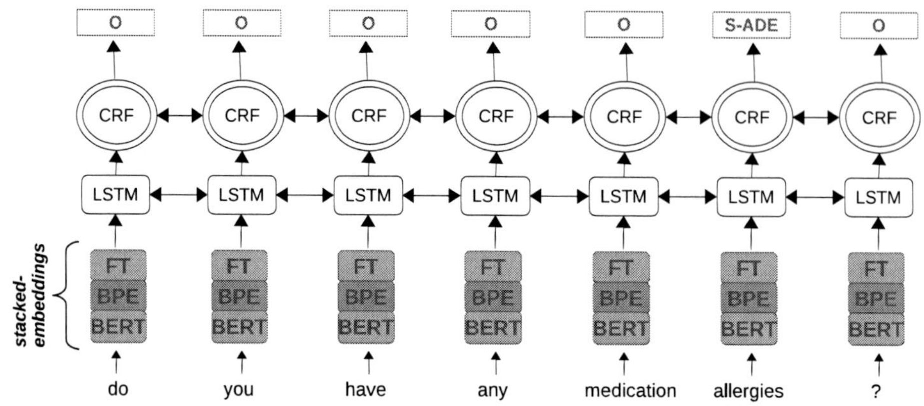

Figure 1: System architecture for NER task, consisting of BiLSTM-CRF with stacked heterogeneous embeddings. Here, *FT*: fastText embedding vector; *BPE*: Byte-Pair embedding vector; *BERT*: BERT embedding vector; *S_ADE*: S_Adverse Drug Effect.

gated various machine learning algorithms like *logistic regression*, *SVM* and *neural network* with various word and sentence embeddings. In *ADE tweet classification* (Task 1a) our submission (team:MIC-NLP) scored F1-score of 0.46, approximately 2 points higher than the arithmetic median of participating submissions; in case of *tweet classification* (task 7a) our system achieved the F1-score of 0.90 which is 5 points higher than the arithmetic median of all submissions.

Task	Train	Dev
Sentence Counts		
Task 1b	34142	1775
Task 7b	14755	4959
Task 1b Entities		
ADE	1713	87
Task 7b Entities		
PROFESION	1597	566
SITUACION_LABORAL	264	85
ACTIVIDAD	45	16
FIGURATIVA	16	8

Table 1: Dataset statistics for NER.

3 Methodology

In the following sections we discuss our proposed model for named entity recognition and text classification.

3.1 Named Entity Recognition

Figure 1 describes the architecture of our model, where we design a sequence tagger to extract entities. The architecture of our model is a standard BiLSTM-CRF (Lample et al., 2016) model with stacked heterogeneous embeddings and linguistic features as input. The stacked embeddings consists of Byte-Pair subword embeddings (Heinzerling and Strube, 2018), fastText subword embeddings (Bojanowski et al., 2017) and contextualized word embeddings (Devlin et al., 2019; Liu et al., 2019). The linguistic features include POS, capitalization features and orthographic features.

3.2 Text Classification

We explored traditional machine learning algorithms like logistic regression, SVM and neural network based architecture with various word and sentence embeddings for text classification. The SVM was trained with Radial Basis Function (RBF) Kernel with the value of penalty parameter C determined by grid search for each dataset. Our best model was a Neural Network with contextualized embeddings (Devlin et al., 2019; Liu et al., 2019). Since both datasets (Task 1a and Task 7a) were highly imbalanced, we employed higher class weights for minority classes to train the final models.

3.3 Ensemble Strategy

Bagging is a useful technique to reduce the variance of the learning algorithm without impacting bias. We employed a variant of Bagging (Breiman, 1996) such that every data point in the training set is part of the development set at least once and vice versa. We created three data folds and trained the model using optimal configuration on each fold, inference on the test set involves majority voting among the

Hyper-parameter	Value
NER	
learning rate	0.1
optimizer	SGD
hidden size	256
POS dimensions	50
Ortho dimension	50
batch size	32
epochs	150
Text Classification	
kernel	RBF
class-weights	10.0
learning rate	0.00003
batch size	16
epochs	10

Table 2: Hyper parameter settings for NER and Text classification.

three trained models.

For NER, we perform majority voting at the token level for each test data point. In cases when voting results in a tie, we take the prediction of the confident model, we treat the model trained on original data split as the confident model. In the case of an ensemble for text classification, we followed the straight forward approach of majority voting at sentence level for each test data point.

4 Experiments and Results

4.1 Dataset and Experimental Setup

Data: We employed bagging (discussed in section 3.3) to split the annotated corpus into 3-folds. For ADE span detection (Task 1b) and Profession span detection (Task 7b) we perform sentence splitting, word tokenization, computing orthographic features and POS tagging. We do not perform any pre-processing for ADE classification (Task 1a) and Tweet classification (Task 7a).

ADE Classification (Task 1a): The dataset consists of tweets in the English language and the task is to detect tweets containing adverse drug effect. The dataset contains two classes, *ADE* and *NoADE*. The dataset is highly imbalanced with only 1235 tweets of type ADE out of total 17385 tweets in the train set.

ADE Span Detection (Task 1b): The dataset consists of only one entity type *ADE*. The train set contains 1717 entity mentions of *ADE* (see Table

	Features	Task 1b P/R/F1	Task 7b P/R/F1
r1	*glove*	.5/.18/.26	-
r2	*fastText*	.89/.28/.43	.84/.64/.73
r3	*fastText + Char*	.64/.28/.39	.83/.67/.74
r4	*fastText + BytePair*	.62/.34/.44	.82/.74/.78
r5	*BERT*	.68/.35/.46	.84/.76/.80
r6	*BERT + fastText + BytePair*	**.61/.52/.56** Fold=2	**.86/.77/.81** Fold=2
r7	*BERT + fastText + BytePair*	.80/.21/.34 Fold=3	.85/.79/.82 Fold=3
r8	*BERT + fastText + BytePair*	.77/.37/.50	.84/.78/.81

Table 3: Scores on dev set using different features for *BiLSTM-CRF* on *Task 1b* and *Task 7b*.

1).

Profession Classification (Task 7a): The dataset consists of tweets in the Spanish language and the task is to detect tweets containing mention of profession/occupation. The dataset contains two classes. The dataset is highly imbalanced with only 1393 tweets containing a positive mention out of 6000 tweets.

Profession Span Detection (Task 7b): The dataset consists of four entity types with few mentions of type *FIGURATIVA* as shown in Table 1. Entities of type ACTIVIDAD and FIGURATIVA are ignored in the evaluation of this shared task but we still treat them as regular entities.

Experimental Setup: We found contextualized embeddings to be very helpful in identifying entities and text classification; all our experiments used pre-trained contextualized embeddings. We employ *RoBERTa* (Gururangan et al., 2020) for Task 1a and Task 1b; we use multi-lingual BERT (Devlin et al., 2019) for Task 7a and Spanish BERT (Cañete et al., 2020) for Task 7b. We do not finetune embeddings in our experiments. We don't employ any strategy for handling imbalanced classes for NER but have used class weighting by a factor of 10 for all positive classes for text classification. Table 2 lists the best configuration of hyperparameters for all the tasks.

4.2 Results on Development Set

We perform various experiments to investigate the impact of features on performance on the development set.

NER: Table 3 shows the score on the development set for Task 1b and Task 7b. Observe that fastText embeddings (row r2) outperform glove embeddings (row r1) for Task 1b. Subsequently, fastText embeddings with BytePair embeddings (row r4) provide an improvement over only fast-

	Features	Task 1a P/R/F1	Task 7a P/R/F1
r1	*logisticReg + fastTextSentEmb*	.33/.83/.47	.38/.95/.55
r2	*logisticReg + BERTSentEmb*	.34/.81/.48	.41/.83/.55
r3	*logisticReg + BERTWordEmbSum*	.45/.86/.59	.45/.86/.59
r4	*SVM + fastTextSentEmb*	.53/.66/.59	.71/.67/.69
r5	*SVM + BERTSentEmb*	.36/.86/.51	.49/.66/.56
r6	*SVM + BERTWordEmbSum*	.44/.90/.59	.61/.64/.63
r7	*NeuralNetwork + Glove*	.51/.63/.56	.64/.59/.61
r8	*NeuralNetwork + BERT*	**.77/.72/.74**	**.95/.85/.90**
r9		**Fold=2**	**Fold=2**
r10	*NeuralNetwork + BERT*	**.79/.66/.72**	**.89/.91/.90**
r11		**Fold=3**	**Fold=3**
r12	*NeuralNetwork + BERT*	**0.8/.65/.72**	**.93/.84/.88**

Table 4: Scores on dev set using different features on *Task 1a* and *Task 7a*.

Text (row r2) and the combination of fastText with Character embeddings (row r3). The contextualized embeddings (row r5) provide an improvement over the combination of fastText with BytePair embeddings. In row r6, we employ BERT, fastText and BytePair embeddings in a stacked format leading to the best f1-score for both Task 1b and Task 7b.

Text Classification: Table 4 shows the score on the development set for Task 1a and Task 7a. Observe that BERTSentEmb provides improvement over fastTextSentEmb for both logistic regression and SVM. Similarly, BERTWordEmbSum further improves BERTSentEmb. BERTSentEmb uses BERT's *CLS* representation whereas BERTWordEmbSum is computed by average of the token-wise embeddings of pre-trained BERT as discussed in Rogers et al.. Neural Network with BERT achieves the best result for both datasets.

4.3 Results on Test Set

Table 5 shows the comparison of our submissions with the arithmetic median of the participating teams for all the tasks. Our submissions achieve the overall best F1-score than the arithmetic median for all the tasks showing compelling advantage. For Task 1a, the precision of our system is lower than the arithmetic median but this is compensated by the improvement in recall. For all the tasks, the precision is higher than the recall but overall precision and recall are balanced.

5 Conclusion

In this paper, we described our system with which we participate in Task 1(Adverse Drug Effect Classification and Extraction) and Task 7 (Identification of professions and occupations in Spanish Tweets) in the SMM4H Shared Task 2021. Our NER system employed stacked heterogeneous em-

	Tasks	Arithmetic Median *P/R/F1*	MIC-NLP *P/R/F1*
r1	*Task 1a*	**.50**/.40/.44	.47/**.45**/**.46**
r2	*Task 1b*	.49/.45/.42	**.55**/**.45**/**.50**
r3	*Task 7a*	.91/.85/.85	**.94**/**.85**/**.90**
r4	*Task 7b*	.84/.72/.76	**.85**/**.79**/**.82**

Table 5: Comparison of our system (team:MIC-NLP) with the arithmetic median of the participating teams. Scores on test set for Task 1a, Task 1b, Task 7a and Task 7b.

beddings to extract entities in English and Spanish text. Our NER system demonstrates a competitive performance with F1-score of 0.50 and 0.82 on ADE Span Detection (Task 1b) and Profession/Occupation span detection (Task 7b) respectively. Our text classification system employed contextualized embeddings with Neural Network as a classifier to achieve a competitive performance with F1-score of 0.46 and 0.90 on ADE Classification (Task 1a) and Profession/Occupation classification (Task 7a) respectively. In future, we would like to improve error analysis to further enhance our NER and text classification models.

Acknowledgment

This research was supported by Bundesministerium für Wirtschaft und Energie (bmwi.de), grant 01MD19003E (PLASS, plass.io) at Technology - Siemens AG, Munich Germany.

References

Cosme Adrover, Todd J. Bodnar, Z. Huang, A. Telenti, and M. Salathé. 2015. Identifying adverse effects of hiv drug treatment and associated sentiments using twitter. volume 1.

Piotr Bojanowski, Edouard Grave, Armand Joulin, and Tomás Mikolov. 2017. Enriching word vectors with subword information. volume 5, pages 135–146.

Leo Breiman. 1996. Bagging predictors. *Machine Learning*, 24(2):123–140.

José Cañete, Gabriel Chaperon, Rodrigo Fuentes, Jou-Hui Ho, Hojin Kang, and Jorge Pérez. 2020. Spanish pre-trained bert model and evaluation data. In *PML4DC at ICLR 2020*.

Jacob Devlin, Ming-Wei Chang, Kenton Lee, and Kristina Toutanova. 2019. BERT: Pre-training of deep bidirectional transformers for language understanding. In *Proceedings of the 2019 Conference of the North American Chapter of the Association for Computational Linguistics: Human Language Technologies, Volume 1 (Long and Short Papers)*, pages 4171–4186, Minneapolis, Minnesota. Association for Computational Linguistics.

Suchin Gururangan, Ana Marasovic, Swabha Swayamdipta, Kyle Lo, Iz Beltagy, Doug Downey, and Noah A. Smith. 2020. Don't stop pretraining: Adapt language models to domains and tasks. In *Proceedings of the 58th Annual Meeting of the Association for Computational Linguistics, ACL 2020, Online, July 5-10, 2020*, pages 8342–8360. Association for Computational Linguistics.

Benjamin Heinzerling and Michael Strube. 2018. BPEmb: Tokenization-free Pre-trained Subword Embeddings in 275 Languages. In *Proceedings of the Eleventh International Conference on Language Resources and Evaluation (LREC 2018)*, Miyazaki, Japan. European Language Resources Association (ELRA).

Keyuan Jiang and Yujing Zheng. 2013. Mining twitter data for potential drug effects. In *ADMA*.

Guillaume Lample, Miguel Ballesteros, Sandeep Subramanian, Kazuya Kawakami, and Chris Dyer. 2016. Neural architectures for named entity recognition. In *Proceedings of the 2016 Conference of the North American Chapter of the Association for Computational Linguistics: Human Language Technologies*, pages 260–270, San Diego, California. Association for Computational Linguistics.

Yinhan Liu, Myle Ott, Naman Goyal, Jingfei Du, Mandar Joshi, Danqi Chen, Omer Levy, Mike Lewis, Luke Zettlemoyer, and Veselin Stoyanov. 2019. Roberta: A robustly optimized BERT pretraining approach. volume abs/1907.11692.

Arjun Magge, Ari Klein, Ivan Flores, Ilseyar Alimova, Mohammed Ali Al-garadi, Antonio Miranda-Escalada, Zulfat Miftahutdinov, Eulàlia Farré-Maduell, Salvador Lima López, Juan M Banda, Karen O'Connor, Abeed Sarker, Elena Tutubalina, Martin Krallinger, Davy Weissenbacher, and Graciela Gonzalez-Hernandez. 2021. Overview of the sixth social media mining for health applications (# smm4h) shared tasks at naacl 2021. In *Proceedings of the Sixth Social Media Mining for Health Applications Workshop & Shared Task*.

Antonio Miranda-Escalada, Eulàlia Farré-Maduell, Salvador Lima López, Vicent Briva-Iglesias, Marvin Agüero-Torales, Luis Gascó-Sánchez, and Martin Krallinger. 2021. The profner shared task on automatic recognition of professions and occupation mentions in social media: systems, evaluation, guidelines, embeddings and corpora. In *Proceedings of the Sixth Social Media Mining for Health Applications Workshop & Shared Task*.

Takeshi Onishi, Davy Weissenbacher, Ari Klein, Karen O'Connor, and Graciela Gonzalez-Hernandez. 2018. Dealing with medication non-adherence expressions in Twitter. In *Proceedings of the 2018 EMNLP Workshop SMM4H: The 3rd Social Media Mining for Health Applications Workshop & Shared Task*, pages 32–33, Brussels, Belgium. Association for Computational Linguistics.

Anna Rogers, Olga Kovaleva, and Anna Rumshisky. 2020. A primer in bertology: What we know about how BERT works. *Trans. Assoc. Comput. Linguistics*, 8:842–866.

BERT based Adverse Drug Effect Tweet Classification

Tanay Kayastha
IIT Bombay, India
kayasthatanay@gmail.com

Pranjal Gupta
BITS Pilani - Hyderabad, India
pranjalgupta2199@gmail.com

Pushpak Bhattacharyya
IIT Bombay, India
pb@cse.iitb.ac.in

Abstract

This paper describes models developed for the Social Media Mining for Health (SMM4H) 2021 shared tasks (Magge et al., 2021). Our team participated in the first subtask that classifies tweets with Adverse Drug Effect (ADE) mentions. Our best performing model utilizes BERTweet followed by a single layer of BiLSTM. The system achieves an F-score of 0.45 on the test set without using any supplementary resources such as Part-of-Speech tags, dependency tags, or knowledge from medical dictionaries.

1 Introduction

In this effort, we focus on detecting tweets that have ADE mentions as a part of the Social Media Mining for Health (#SMM4H) - 2021 shared tasks (Magge et al., 2021). Organizers of SMM4H Task 1 provided datasets of English tweets with binary annotations of 1 and 0 indicating the presence or absence of ADE mentions in the tweet. We develop a robust system against the class imbalance problem in the dataset that classifies tweets containing at least one ADE mention. We also validate the importance of emojis and hashtags in ADE classification empirically.

2 Data

2.1 Dataset

The dataset consists of a training set (18,000 tweets), validation set (953 tweets), and test set (10,000 tweets). The dataset is highly imbalanced, with only 7% of the tweets containing ADE mentions. We tackle this challenge using sampling and per-class penalties in the objective function.

2.2 Preprocessing

We performed following preprocessing on the dataset:

1. Replace emoji with its text string (for example, ':)' with 'slightly smiling face')
2. Strip '#' from hashtags in tweets
3. Drop user-mentions and URLs
4. Lowercase all words

We used `emoji`[1] package to translate emoji to text string.

3 Method

We explore three BERT-based models for classification: (i) BERT (Devlin et al., 2019), (ii) RoBERTa (Liu et al., 2019), and (iii) BERTweet (Nguyen et al., 2020). We pass the input through our BERT-based models to get token representations. To compute the sentence representations, we consider two cases - i) `[CLS]` token (fine-tuning) ii) we pass token representations without `[CLS]` and `[SEP]` through a single layer BiLSTM and concatenate the forward and backward context. The sentence representation is passed through a fully connected neural network layer followed by a sigmoid activation to predict probabilities.

To tackle class imbalance, we experiment with oversampling, undersampling, and addition of per-class penalties in the objective function. For oversampling approach, we randomly sampled positive examples with replacement until each class contained 10,000 tweets. For the undersampling approach, we randomly sample negative examples to create a balanced training dataset.

4 Experiments

For the classification task, each BERT model is trained for 10 epochs with a learning rate of $1 * 10^{-5}$ using Adam optimizer (Kingma and Ba,

[1] https://pypi.org/project/emoji/

Proceedings of the Sixth Social Media Mining for Health Workshop 2021, pages 88–90
June 10, 2021.

Name	Validation set		
	P	**R**	**F1**
BERT_{base} - Fine Tune	0.697	0.708	0.702
BERT_{base} - unweighted	0.742	0.708	0.724
BERT_{base}	0.77	0.723	0.746
RoBERTa	0.845	0.754	0.797
RoBERTa_{over}	0.637	**0.892**	0.743
RoBERTa_{under}	0.659	0.862	0.747
BERTweet_{raw}	**0.864**	0.784	0.823
BERTweet	0.812	0.862	**0.836**

Table 1: Task1a results on Validation set

2017). We set the batch size to 32 and the maximum sequence length to 128. To tackle class imbalance, we add weights to the standard cross-entropy loss. We set weights as 0.7 and 0.3 for ADE and NoADE classes, respectively. We utilize PyTorch[2] implementation of BERT for training. We train RoBERTa_{over}, RoBERTa_{under} and BERT_{base} - unweighted, using standard *unweighted* cross-entropy loss. We conduct model selection for every 200 steps against the validation set using the F1-score of the ADE class for comparison.

5 Discussion

It is evident from Table 1 that BERT_{base} outperforms BERT_{base}-Fine Tune, and validates that the use of BiLSTM layer on top of BERT improves both precision and recall. Table 1 also shows that use of per-class penalties in the objective function (BERT_{base}) results in better performance as compared to the model with *unweighted* objective function (BERT_{base} - unweighted).

Table 2 shows that retaining emoji and hashtags in tweets help in achieving better performance on BERT_{base} as against excluding those.

Table 1 shows that RoBERTa outperformed BERT_{base} in all the evaluation metrics. However, RoBERTa_{over} and RoBERTa_{under} gave results comparable to BERT_{base}. The results show that the ADE class's oversampling and the NoADE class's undersampling did not handle the class imbalance problem well. Hence, we resort to adding class-weights in our objective function.

BERTweet outperforms BERTweet_{raw}, which uses preprocessing techniques described in (Nguyen et al., 2020). Our preprocessing steps are inspired by (Nguyen et al., 2020) with the only difference being that we remove all user mentions and web/URL links from the tweet. We empirically validate our intuition that the user mentions, web links act as noise in the text and do not provide any valuable information needed for the classification task.

Table 3 shows the performance of BERTweet on the test set. Our model's performance is relatively poor on the Test set compared to the validation set, which can be attributed to overfitting. This overfitting can be reduced by adding dropout in the model. Table 3 shows the performance of BERTweet on the Test set in the post-evaluation phase after the addition of dropout to the BiLSTM layers.

Model: $\textbf{BERT}_{base}$	**P**	**R**	**F1**
retain hashtag	**0.80**	0.677	0.733
retain emoji	0.671	**0.754**	0.71
retain hashtag and emoji	0.77	0.723	**0.746**

Table 2: Results of BERT_{base} trained with different preprocessing applied both to training and validation set

Model: **BERTweet**	**P**	**R**	**F1**
Evaluation	0.523	0.409	0.46
Post-Evaluation	**0.538**	**0.451**	**0.491**

Table 3: Results of BERTweet on #SMM4H - 2021 Task 1a Test set

6 Conclusion

In this work, we explore an application of BERT to the task of binary classification on English Tweets. We validate that use of per-class penalties in the objective function helped in overcoming the class imbalance problem. We have empirically evaluated differently tuned model versions and preprocessing methods against F1-score for the "ADE" class. Experiments have shown that our model has achieved an F1-score of 0.46, precision of 0.523, and recall of 0.409 on the test set.

The future directions would be to evaluate the potential of supplementary resources in our model, such as Part-of-Speech Tags, Dependency Tags, knowledge from medical dictionaries (such as MedDRA).

References

Jacob Devlin, Ming-Wei Chang, Kenton Lee, and Kristina Toutanova. 2019. Bert: Pre-training of deep

[2]https://huggingface.co/transformers/model_doc/bert.html

bidirectional transformers for language understanding.

Diederik P. Kingma and Jimmy Ba. 2017. Adam: A method for stochastic optimization.

Yinhan Liu, Myle Ott, Naman Goyal, Jingfei Du, Mandar Joshi, Danqi Chen, Omer Levy, Mike Lewis, Luke Zettlemoyer, and Veselin Stoyanov. 2019. Roberta: A robustly optimized bert pretraining approach.

Arjun Magge, Ari Klein, Ivan Flores, Ilseyar Alimova, Mohammed Ali Al-garadi, Antonio Miranda-Escalada, Zulfat Miftahutdinov, Eulàlia Farré-Maduell, Salvador Lima López, Juan M Banda, Karen O'Connor, Abeed Sarker, Elena Tutubalina, Martin Krallinger, Davy Weissenbacher, and Graciela Gonzalez-Hernandez. 2021. Overview of the sixth social media mining for health applications (# smm4h) shared tasks at naacl 2021. In *Proceedings of the Sixth Social Media Mining for Health Applications Workshop & Shared Task.*

Dat Quoc Nguyen, Thanh Vu, and Anh Tuan Nguyen. 2020. BERTweet: A pre-trained language model for English Tweets. In *Proceedings of the 2020 Conference on Empirical Methods in Natural Language Processing: System Demonstrations*, pages 9–14.

A Joint Training Approach to Tweet Classification and Adverse Effect Extraction and Normalization for SMM4H 2021

Mohab Elkaref
IBM Research Europe
Daresbury, United Kingdom
mohab.elkaref@ibm.com

Lamiece Hassan
University of Manchester
Manchester, United Kingdom
lamiece.hassan@manchester.ac.uk

Abstract

In this work we describe our submissions to the Social Media Mining for Health (SMM4H) 2021 Shared Task (Magge et al., 2021). We investigated the effectiveness of a joint training approach to Task 1, specifically classification, extraction and normalization of Adverse Drug Effect (ADE) mentions in English tweets. Our approach performed well on the normalization task, achieving an above average f1 score of 24%, but less so on classification and extraction, with f1 scores of 22% and 37% respectively. Our experiments also showed that a larger dataset with more negative results led to stronger results than a smaller more balanced dataset, even when both datasets have the same positive examples. Finally we also submitted a tuned BERT model for Task 6: Classification of Covid-19 tweets containing symptoms, which achieved an above average f1 score of 96%.

1 Introduction

Social media platforms such as Twitter are regarded as potentially valuable tools for monitoring public health, including identifying ADEs to aid pharmacovigilance efforts. They do however pose a challenge due to the relative scarcity of relevant tweets in addition to a more fluid use of language, creating a further challenge of identifying and classifying specific instances of health-related issues. In this year's task as well as previous SMM4H runs (Klein et al., 2020) a distinction is made between classification, extraction, and normalization. This is atypical of NER systems, and many other NER datasets present their datasets, and are consequently solved in a joint approach.

Gattepaille (2020) showed that simply tuning a base BERT (Devlin et al., 2019) model could achieve strong results, even beating ensemble methods that rely on tranformers pretrained on more academic texts such as SciBERT (Beltagy et al., 2019), BioBERT (Lee et al., 2020) or ensembles of them, while approaching the performance of BERT models specifically pretrained on noisy health-related comments (Miftahutdinov et al., 2020).

2 Methods

2.1 Pre-processing

Despite the noisy nature of Twitter data, for Task 1 we attempted to keep any pre-processing to a minimum. This was motivated by the presence of spans within usernames and hashtags, in addition to overlapping spans and spans that included preceding or trailing white-spaces. For training and validation data we ignored overlapping and nested spans and chose the longest span as the training/tuning example.

We also compiled a list of characters used in the training data for use in creating character embeddings. This was not limited to alpha-numeric characters, but also included emojis, punctuation, and non-Latin characters. We then removed any character appearing less than 20 times[1] in the training set, and a special UNK character embedding was added. Additionally for the training, validation, and testing data we tokenized the tweets and obtained part-of-speech tags using the default English model for the Stanza (Qi et al., 2020) pipeline.

Our training set was supplemented with the CSIRO Adverse Drug Event Corpus(CADEC) (Karimi et al., 2015) and was processed in the same manner as above.

For Task 6 no pre-processing was done.

2.2 Task 1 Model

Word Representation The BERT vectors produced for each tweet are not necessarily aligned with the tokens produced by the Stanza tokenizer. For this reason we additionally compile a sub-word token map to construct word embeddings from the token embeddings produced by our BERT model

[1]This threshold was arrived at through trial and error.

Proceedings of the Sixth Social Media Mining for Health Workshop 2021, pages 91–94
June 10, 2021. ©2021 Association for Computational Linguistics

(excluding the [CLS] vector). The final word embedding is a summation of the component vectors.

POS Tags & Char-LSTM We use randomly initialized trainable embeddings for universal POS (UPOS) tags predicted by the Stanza pos-tagger. For each word we also use a 1-layer LSTM to produce an additional representation. The input to this LSTM would be the embeddings for each character in a word in order of appearance. This is intended to capture both the recurring patterns indicating prefixes/suffixes and to also learn to disregard repeated letters and misspellings so as to overcome the noisiness of the data.

Bi-LSTM Hidden Layer While BERT is itself a bi-directional context-aware representation of a given sentence, we experimented with the addition of a bidirectional Lstm (Bi-LSTM) layer in order to incorporate the additional pos tag and char-LSTM embeddings, and model the interactions between them across the whole context of a tweet.

ADE Identification Subtask 1(a) requires simple classification as ADE or NoADE and so we simply used the [CLS] vector output from our BERT model as input to a softmax layer with two nodes.

ADE Extraction & Normalization Task 1(b) and 1(c) were approached jointly. The training data was reformulated a BIO labelling scheme that incoporates associated MedDRA tags, as is common for other NER tasks. Thus the final classification layer for both tags is a softmax with $(\{B, I\} \times MedDRA\,Tags + \{O\})$-nodes. We use a greedy approach to obtain the final tweet classification and token classification from the corresponding softmax layers. The spans are determined based on the longest uninterrupted sequence of tokens receiving the same *normalization* tag. Interruptions in this context mean classified as either O or B-⋆. Additionally, uninterrupted spans consisting only of I-⋆ but having the same normalization tag are considered valid spans. Thus, the following two sequences ([O,O,B-1234,I-1234,O] and [O,O,I-1234,I-1234,O]) translate to the same final span.

2.3 Task 6 Model

Task 6 proved to be a substantially easier challenge than Subtask 1(a), as can be seen in Subsection 3.2. Our approach was to simply tune a BERT model, with the [CLS] vector being used as input to a softmax classification layer.

Parameter	
Task 1	
POS embedding dimension	8
Character embedding dimension	16
Character LSTM dimension	8
Bi-LSTM hidden layer dimension	256
Training Epochs	50
Mini-batch size	32
Update Strategy	Adam
Learning Rate	$1 \times 10^{-4}/2 \times 10^{-5}$
Task 6	
Training Epochs	10
Mini-batch size	8
Update Strategy	Adam
Learning Rate	$1 \times 10^{-5}/2 \times 10^{-5}$

Table 1: Training parameters for Tasks 1 & 6

3 Experiments & Results

We implemented our models using the PyTorch (Paszke et al., 2019) framework, and for the core BERT model we used the pretrained bert-base model from the Huggingface transformers (Wolf et al., 2020) library. For both tasks we optimize parameters using Adam (Kingma and Ba, 2014). We experiment with different learning rates but keep default parameters for β_1, β_2, and ϵ.

3.1 Task 1

One of the largest challenges of Task 1 is the huge imbalance of tweets containing ADEs vs tweets that do not. This is demonstrated in Table 3 where just over 7% of tweets in both training and validation sets contain ADEs. In contrast, the CADEC dataset has ≈ 37% of examples with ADEs. To explore the effect of this distribution we constructed two training sets. The first is a dataset containing all the CADEC data in addition to training data tweets containing ADEs. This results in a dataset with ≈ 46% of examples with ADEs, which we will refer to as the *Partial datatset* going forward. The second dataset we use for training is all of the task training data and the whole CADEC dataset, which we will be referring to as the *Full dataset*, with the proportion of ADE examples being ≈ 16%.

We train the model jointly over all three subtasks, minimizing over the sum of negative log likelihood losses ($L_{SUM} = L_{DET} + L_{NER}$) for both the classification ($L_{DET} = -\sum_i^N \sum_c^{C_{DET}} y_{ic} log(\hat{y}_{ic})$) and extraction & normalization ($L_{NER} = -\sum_i^N \sum_c^{C_{NER}} y_{ic} log(\hat{y}_{ic})$) layers. Where N is the total number of minibatches, C_{DET} and C_{NER} are the classes for classification and extraction & normalization respectively, and y_* and $\hat{y}_*$ are the target and predicted classes.

Train dataset	Classification			Extraction			Normalization		
Target dataset	f1	p	r	f1	p	r	f1	p	r
Partial dataset									
Validation (1×10^{-4})	14.9	8.0	100.0	10.5	6.1	37.9	19.1	11.1	69.0
Validation (2×10^{-5})	14.8	8.0	100.0	9.3	5.5	29.9	18.2	10.8	58.6
Full dataset									
Validation	70.1	78.8	63.1	26.9	27.4	26.4	50.3	51.2	49.4
Test	22.0	35.9	16.4	37.0	58.0	27.5	24.0	37.1	17.8
Median of all submissions									
Test	44.0	50.5	40.9	42.0	49.3	45.8	22.0	23.1	21.8

Table 2: Task 1 Experimental Results.

Dataset	Total tweets	ADE tweets
CADEC	7597	2853
Training data	17358	1235
Validation data	915	65

Table 3: Task 1 dataset statistics.

α	Validation			Test		
	f1	p	r	f1	p	r
1×10^{-5}	98.3	98.2	98.3	94.0	94.1	94.1
2×10^{-5}	98.6	98.6	98.6	94.0	93.7	93.7
Median of all submissions				93.0	93.2	93.2

Table 4: Task 6 Experimental Results.

Our experiments on the *partial datasets* yielded weak results, with only a slight improvement when using a learning rate of 1×10^{-4} over 2×10^{-5}. Training on the *full dataset* with a learning rate of 2×10^{-5} produced far stronger results, with the f1 score for tweet classification increasing to 70.1% from 14.9% on the validation set, and to 26.9% from 10.5% for span extraction, and finally to 50.4% from 19.1% for span normalization. Training our model with a learning rate of 1×10^{-4} yielded unusable results and an unstable model, which suggests that this is too high a learning rate for larger datasets. It is interesting to note that while training on the full datatset dramatically improved f1 scores for all three subtasks, there was a general drop in recall and an increase in precision. This suggests that the model trained on the partial dataset was far more likely to produce false positives, and was unable to recognize the absence of ADEs despite negative examples constituting $\approx$ 53% of examples. The results of our experiments are summarized in Table 2.

Our final submission was trained on the full dataset and showed a similar pattern on the Test set producing better precision, beating the arithmetic mean of all submissions for extraction and normalization, but showed worse recall for all three subtasks. This resulted in the model only achieving an above average f1 score on subtask 1(c).

3.2 Task 6

Our approach to Task 6 is essentially the same as that for subtask 1(a), but with a smaller, more balanced dataset. We experiment with two learning rates, 1×10^{-5} and 2×10^{-5}, and minimize over a negative log likelihood loss $L = -\sum_i^N \sum_c^C y_{ic} log(\hat{y}_{ic})$.

The resulting models produced strong results, as shown in Table 4, with close validation f1 scores (98.6% and 98.3%). We used classifications by both models as our final submission, and both beat the median of all submissions with an f1 score of 94% for both models.

4 Conclusion

In this work we explored the efficacy of jointly training a BERT model to jointly learn to perform classification, extraction, and normalization of ADE in tweets provided for Task 1 in SMMH 2021 Shared Task. While this approach did not produce classification and extraction above the median submission, it did achieve a normalization score that is. Additionally our experiments show that the seemingly lopsided ratio of tweets with/without ADEs resulted in stronger performance than a more "balanced" dataset. Finally, we showed that tuning a BERT model produces very strong results on Task 6, in classifying tweets related to Covid-19.

Acknowledgements

This work was part of STFC Hartree Centre's Innovation Return on Research programme, Contract Number: 4070116091. LH was funded via the Medical Research Council (Ref: MR/S004025/1).

References

Iz Beltagy, Kyle Lo, and Arman Cohan. 2019. Scibert: A pretrained language model for scientific text. *arXiv preprint arXiv:1903.10676*.

Jacob Devlin, Ming-Wei Chang, Kenton Lee, and Kristina Toutanova. 2019. BERT: Pre-training of deep bidirectional transformers for language understanding. In *Proceedings of the 2019 Conference of the North American Chapter of the Association for Computational Linguistics: Human Language Technologies, Volume 1 (Long and Short Papers)*, pages 4171–4186, Minneapolis, Minnesota. Association for Computational Linguistics.

Lucie Gattepaille. 2020. How far can we go with just out-of-the-box bert models? In *Proceedings of the Fifth Social Media Mining for Health Applications Workshop & Shared Task*, pages 95–100.

Sarvnaz Karimi, Alejandro Metke-Jimenez, Madonna Kemp, and Chen Wang. 2015. Cadec: A corpus of adverse drug event annotations. *Journal of biomedical informatics*, 55:73–81.

Diederik Kingma and Jimmy Ba. 2014. Adam: A method for stochastic optimization. *arXiv preprint arXiv:1412.6980*.

Ari Klein, Ilseyar Alimova, Ivan Flores, Arjun Magge, Zulfat Miftahutdinov, Anne-Lyse Minard, Karen O'Connor, Abeed Sarker, Elena Tutubalina, Davy Weissenbacher, and Graciela Gonzalez-Hernandez. 2020. Overview of the fifth social media mining for health applications (#SMM4H) shared tasks at COLING 2020. In *Proceedings of the Fifth Social Media Mining for Health Applications Workshop & Shared Task*, pages 27–36, Barcelona, Spain (Online). Association for Computational Linguistics.

Jinhyuk Lee, Wonjin Yoon, Sungdong Kim, Donghyeon Kim, Sunkyu Kim, Chan Ho So, and Jaewoo Kang. 2020. Biobert: a pre-trained biomedical language representation model for biomedical text mining. *Bioinformatics*, 36(4):1234–1240.

Arjun Magge, Ari Klein, Ivan Flores, Ilseyar Alimova, Mohammed Ali Al-garadi, Antonio Miranda-Escalada, Zulfat Miftahutdinov, Eulàlia Farré-Maduell, Salvador Lima López, Juan M Banda, Karen O'Connor, Abeed Sarker, Elena Tutubalina, Martin Krallinger, Davy Weissenbacher, and Graciela Gonzalez-Hernandez. 2021. Overview of the sixth social media mining for health applications (# smm4h) shared tasks at naacl 2021. In *Proceedings of the Sixth Social Media Mining for Health Applications Workshop & Shared Task*.

Zulfat Miftahutdinov, Andrey Sakhovskiy, and Elena Tutubalina. 2020. KFU NLP team at SMM4H 2020 tasks: Cross-lingual transfer learning with pretrained language models for drug reactions. In *Proceedings of the Fifth Social Media Mining for Health Applications Workshop & Shared Task*, pages 51–56, Barcelona, Spain (Online). Association for Computational Linguistics.

Adam Paszke, Sam Gross, Francisco Massa, Adam Lerer, James Bradbury, Gregory Chanan, Trevor Killeen, Zeming Lin, Natalia Gimelshein, Luca Antiga, Alban Desmaison, Andreas Kopf, Edward Yang, Zachary DeVito, Martin Raison, Alykhan Tejani, Sasank Chilamkurthy, Benoit Steiner, Lu Fang, Junjie Bai, and Soumith Chintala. 2019. Pytorch: An imperative style, high-performance deep learning library. In H. Wallach, H. Larochelle, A. Beygelzimer, F. d'Alché-Buc, E. Fox, and R. Garnett, editors, *Advances in Neural Information Processing Systems 32*, pages 8024–8035. Curran Associates, Inc.

Peng Qi, Yuhao Zhang, Yuhui Zhang, Jason Bolton, and Christopher D. Manning. 2020. Stanza: A Python natural language processing toolkit for many human languages. In *Proceedings of the 58th Annual Meeting of the Association for Computational Linguistics: System Demonstrations*.

Thomas Wolf, Lysandre Debut, Victor Sanh, Julien Chaumond, Clement Delangue, Anthony Moi, Pierric Cistac, Tim Rault, Remi Louf, Morgan Funtowicz, Joe Davison, Sam Shleifer, Patrick von Platen, Clara Ma, Yacine Jernite, Julien Plu, Canwen Xu, Teven Le Scao, Sylvain Gugger, Mariama Drame, Quentin Lhoest, and Alexander Rush. 2020. Transformers: State-of-the-art natural language processing. In *Proceedings of the 2020 Conference on Empirical Methods in Natural Language Processing: System Demonstrations*, pages 38–45, Online. Association for Computational Linguistics.

Text Augmentation Techniques in Drug Adverse Effect Detection Task

Pavel Blinov
Sber Artificial Intelligence Laboratory / Moscow, Russia
Blinov.P.D@sberbank.ru

Abstract

The paper researches the problem of drug adverse effect detection in texts of social media. We describe the development of such a classification system for Russian tweets. To increase the training dataset we apply a couple of augmentation techniques and analyze their effect in comparison with similar systems presented at 2021 years' SMM4H Workshop.

1 Introduction

Attention-based neural network models significantly move forward performance frontier for a range of Natural Language Processing (NLP) tasks. Pre-trained models with transformer architectures (Devlin et al., 2018; Liu et al., 2019) essentially changed the way itself of approaching an NLP problem. Fine-tuning such models for a specific task typically yields a solid result. But there are still challenging problems even among the simplest binary text classification tasks. For example, for current state-of-the-art NLP methods, it is not an easy task to differentiate drug Adverse Effects (AE) mentions among real indications for use. Especially if the target text comes from informal data sources (see example in Section 2). For several recent years, this problem stays in research focus and is offered as a shared task during the annual Social Media Mining for Health Applications (SMM4H) workshop (Magge et al., 2021). And the second time it was proposed for the Russian language.

The training data size can be crucial for deep learning algorithms generalization hence the performance metrics (Chen and Lin, 2014). This study explores the ways of gaining additional train data. We describe a couple of such techniques (translation and generation) and apply them to increase the training dataset more than 9 times.

2 Data

The SMM4H workshop organizers released *Train* and *Dev* data (user messages from Twitter) along with target labels. The pair of examples (translated from Russian for readability) are listed below:

I finally finished drinking this Tavanik. From which insomnia. ⇒ ***1***

The main symptoms of a lack of thyroxine are just obesity, decreased intelligence, chilliness and insomnia. ⇒ ***0***

Part	Count	Positive ratio, (%)
Train	8,184	9.45
Dev	3,425	8.73
Test	9,095	n/a
Augm_Transl	25,678	9.26
Augm_Gen	51,152	9.89
Total	97,534	n/a

Table 1: Dataset statistics.

Statistics about data parts are shown in Table 1. The *Augm_** rows are additional labeled data[1] (see Section 3 for details).

3 System Description

Data augmentation techniques are well presented in the computer vision field (Shorten and Khoshgoftaar, 2019). Distortion of an input image allows getting an additional data sample. Unfortunately for NLP tasks, there are no simple and effective operations to mine new data samples. Mere word order change or replacement of words often leads to loss or change of text meaning. That is because natural language obeys numerous rules and restrictions. To account for most of these rules and 'correctly' transform a text one needs to rely on a language model.

[1]Available for download at https://disk.yandex.ru/d/BQ-YM8MIsni7VQ

Proceedings of the Sixth Social Media Mining for Health Workshop 2021, pages 95–97
June 10, 2021.

System	Train samples	CV $F_1 \pm F_{std}$	Test Precision	Test Recall	Test F_1
Median			54.9	55.7	51
Real+Augm_*	86,111	57.2±2.5	39.3	59	47
Real+Augm_Transl	34,959	56.6±2.5			
Real	9,282	55.4±2.2			

Table 2: Systems performance metrics, (%).

3.1 Translate Augmentations

Having a long history of research current neural machine translation methods achieve great success in conveying the meaning and keeping text fluency. This allows the implementation of the idea of back translation for text data augmentation (Edunov et al., 2018). Target text translated from a source to destination language then back to the source language, e.g. ru ⇒ en ⇒ ru. Thus the final translation will contain a slightly different sample.

We apply a shortened version of such pipeline (en ⇒ ru) as we had an English dataset from the previous iteration of SMM4H workshop. In such a way we obtain an additional train part (*Augm_Transl*) of 25,678 samples.

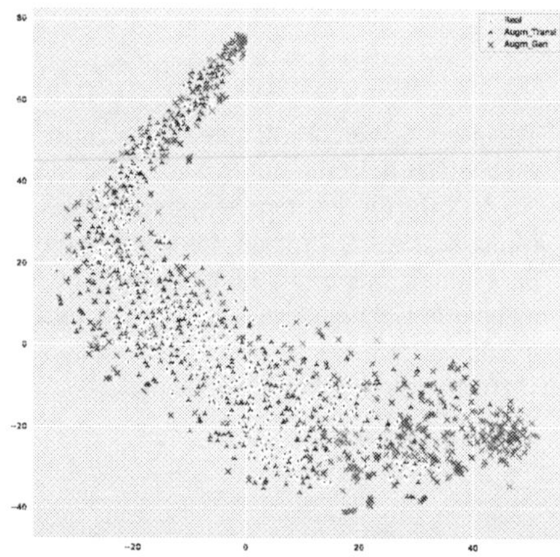

Figure 1: Samples of tweet embeddings from 3 data parts.

3.2 Generation Augmentations

Besides specialized language models for translation, there is the class of Generative Pre-Training models (e.g. GPT-2) (Radford et al., 2019). Such models, trained for a phrase continuation task, could produce surprisingly plausible and coherent text fragments.

Similar to (Blinov, 2020) we adopt and fine-tune the GPT-2 model for the task of Russian tweet generation. Given a couple of random start tokens, the trained model can complete a tweet message. From this model, we retrieved 100k synthetic unlabeled messages and applied our model (Blinov and Avetisian, 2020) for labeling. Finally, only 51,152 samples with high confidence labels were selected, which comprise the *Augm_Gen* part.

3.3 Modeling

To build the final classifier we used the RuBERT (Kuratov and Arkhipov, 2019) model as a base. It was fine-tuned on the mix of augmented and real labeled data with the mean pooling strategy over contextualized set of token embeddings and binary cross-entropy loss function.

More precisely we prepared 5 of such models according to Cross-Validation (CV) split on the concatenation of *Train* and *Dev* parts. Each fold's train data was joined with *Augm_** parts and a model trained for 5 epochs with a batch size of 128 samples and 3×10^{-5} learning rate.

As we required to output binary prediction value each epoch training followed by the threshold optimization procedure. In the end, we selected the best model checkpoint and threshold for each of 5 folds. At the test time, input data processed by 5 models, and their output are binarized. The final label for a sample selected as the most common one.

4 Results and Conclusions

F_1-score toward the positive class (Manning et al., 2008) is the main evaluation metric for this task. Table 2 reports cross-validation and test metrics for a number of our systems. As we keep validation sets intact and increase with additional data only train parts we can compare the metric across systems. The *Real** prefix in a system name corresponds to this year's data (*Train* and *Dev* parts from Table 1).

Although we can see clear CV metric improvement it turned out that it did not convert into test performance. Our best system is inferior to even the median metric across participants' systems, overcoming it only in terms of Recall (by the 3% margin).

We hypothesize that this is because of a significant shift in data distribution. Partially it is confirmed by t-SNE (Maaten and Hinton, 2008) plot of randomly sample tweet embeddings from three data parts (see Figure 1), where synthetically generated messages concentrate on the border of the point cloud.

Thus our experiments reveal that procedures of text data augmentation potentially are an interesting tool for obtaining more data. But the successful practical application of these techniques for the AE detection task requires further research.

References

Pavel Blinov. 2020. Semantic triples verbalization with generative pre-training model. In *Proceedings of the 3rd International Workshop on Natural Language Generation from the Semantic Web (WebNLG+)*, pages 154–158, Dublin, Ireland (Virtual). Association for Computational Linguistics.

Pavel Blinov and Manvel Avetisian. 2020. Transformer models for drug adverse effects detection from tweets. In *Proceedings of the Fifth Social Media Mining for Health Applications Workshop & Shared Task*, pages 110–112, Barcelona, Spain (Online). Association for Computational Linguistics.

Xue-Wen Chen and Xiaotong Lin. 2014. Big data deep learning: challenges and perspectives. *IEEE access*, 2:514–525.

Jacob Devlin, Ming-Wei Chang, Kenton Lee, and Kristina Toutanova. 2018. Bert: Pre-training of deep bidirectional transformers for language understanding. *arXiv preprint arXiv:1810.04805*.

Sergey Edunov, Myle Ott, Michael Auli, and David Grangier. 2018. Understanding back-translation at scale. *arXiv preprint arXiv:1808.09381*.

Yuri Kuratov and Mikhail Arkhipov. 2019. Adaptation of deep bidirectional multilingual transformers for russian language. *Computing Research Repository*, arXiv:1905.07213.

Yinhan Liu, Myle Ott, Naman Goyal, Jingfei Du, Mandar Joshi, Danqi Chen, Omer Levy, Mike Lewis, Luke Zettlemoyer, and Veselin Stoyanov. 2019. Roberta: A robustly optimized BERT pretraining approach. *Computing Research Repository*, arXiv:1907.11692.

Laurens van der Maaten and Geoffrey Hinton. 2008. Visualizing data using t-sne. *Journal of machine learning research*, 9(Nov):2579–2605.

Arjun Magge, Ari Klein, Ivan Flores, Ilseyar Alimova, Mohammed Ali Al-garadi, Antonio Miranda-Escalada, Zulfat Miftahutdinov, Eulàlia Farré-Maduell, Salvador Lima López, Juan M Banda, Karen O'Connor, Abeed Sarker, Elena Tutubalina, Martin Krallinger, Davy Weissenbacher, and Graciela Gonzalez-Hernandez. 2021. Overview of the sixth social media mining for health applications (# smm4h) shared tasks at naacl 2021. In *Proceedings of the Sixth Social Media Mining for Health Applications Workshop & Shared Task*.

Christopher D. Manning, Prabhakar Raghavan, and Hinrich Schütze. 2008. *Introduction to Information Retrieval*. Cambridge University Press, USA.

Alec Radford, Jeffrey Wu, Rewon Child, David Luan, Dario Amodei, and Ilya Sutskever. 2019. Language models are unsupervised multitask learners. *OpenAI blog*, 1(8):9.

Connor Shorten and Taghi M Khoshgoftaar. 2019. A survey on image data augmentation for deep learning. *Journal of Big Data*, 6(1):1–48.

Classification of Tweets Self-reporting Adverse Pregnancy Outcomes and Potential COVID-19 Cases Using RoBERTa Transformers

Lung-Hao Lee, Man-Chen Hung, Chien-Huan Lu,
Chang-Hao Chen, Po-Lei Lee, and Kuo-Kai Shyu
Department of Electrical Engineering, National Central University, Taiwan
Pervasive Artificial Intelligence Research (PAIR) Labs, Taiwan

Abstract

This study describes our proposed model design for SMM4H 2021 shared tasks. We fine-tune the language model of RoBERTa transformers and their connecting classifier to complete the classification tasks of tweets for adverse pregnancy outcomes (Task 4) and potential COVID-19 cases (Task 5). The evaluation metric is F1-score of the positive class for both tasks. For Task 4, our best score of 0.93 exceeded the median score of 0.925. For Task 5, our best of 0.75 exceeded the median score of 0.745.

1 Introduction

The Social Media Mining for Health Application (SMM4H) shared tasks involve natural language processing challenges using social media data for health research. We participated in the SMM4H 2021 Task 4 (Klein et al., 2020a; 2020b), focusing on automatically distinguishing tweets that self-report a personal experience of an adverse pregnancy including miscarriage, stillbirth, preterm birth, low birthweight, and neonatal intensive care (annotated as "1") from those that do not (annotated as "0"). This task is a follow-up to SMM4H 2020 Task 5, which involves three classes of tweets that mention birth defects.

We also participated in SMM4H 2021 Task 5. This new binary classification task involves automatically distinguishing tweets that self-report potential cases of COVID-19 (annotated as "1") from those that do not (annotated as "0"). Potential cases includes those tweets indicate the user or a member of the user's household was denied testing for, was symptomatic of, was directly exposed to presumptive or confirmed COVID-19 cases, or had experiences that pose a higher risk of exposure to COVID-19. Other tweets related to COVID-19 may discuss topics such as testing, symptom, traveling, or social distancing, but do not indicate someone may be infected.

This paper describes the **NCUEE-NLP** (**N**ational **C**entral **U**niversity, Dept. of **E**lectrical **E**ngineering, **N**atural **L**anguage **P**rocessing Lab) system for the SMM4H 2021 Task 4 and Task 5. Our solution explores how to use the RoBERTa transformers (Liu et al., 2019) with involved language models and classifier fine-tuning to predict tweet classes. The evaluation metrics of both tasks are F1-score for the positive class (i.e., tweets annotated as "1"). For Task 4, our best score of 0.93 exceeded the median score of 0.925. For Task 5, we achieved a best score of 0.75 exceeding the median score of 0.745.

The rest of this paper is organized as follows. Section 2 investigates the related studies. Section 3 describes the NCUEE-NLP system for the tweet classification tasks. Section 4 presents the evaluation results and performance comparisons. Conclusions are finally drawn in Section 5.

2 Related Work

Our participated SMM4H 2021 Task 4 is a follow-up to SMM4H 2020 Task 5, which focused on detecting tweets that mention birth defects. A hard-voting ensemble of nine BioBERT-based models was used to achieve a higher macro-averaging recall (Bai and Zhou, 2020). The ELMo word embeddings and data-specific resources were adopted to achieve a higher macro-averaging precision (Bagherzadeh and Bergler, 2020). Ensemble BERT flavors were studied to detect tweets that mention birth defects (Dima et al., 2020). Two-views based CNN-BiGRU networks

Proceedings of the Sixth Social Media Mining for Health Workshop 2021, pages 98–101
June 10, 2021.

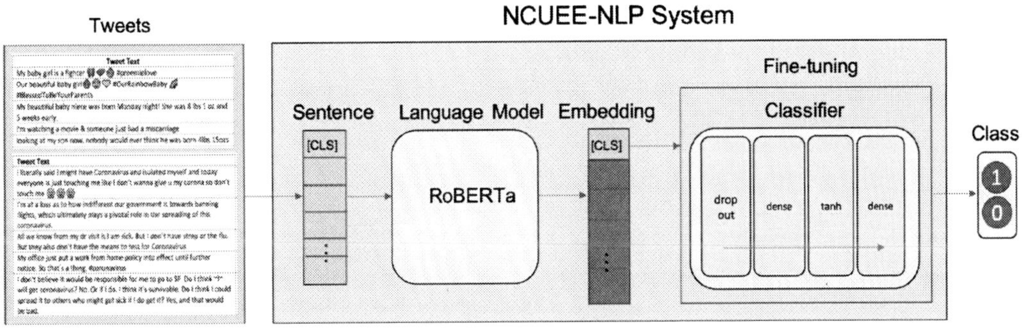

Figure 1: Our NCUEE-NLP system architecture for the SMM4H 2021 Task 4 and Task 5.

RoBERTa Transformers	Fine-Tuning		Validation Set			Test Set		
	LM	Classifier	Precision	Recall	F1-score	Precision	Recall	F1-score
	No	Yes	0.9253	0.9338	0.9296	0.9235	0.9248	0.92
	Yes	Yes	0.9141	0.9475	**0.9305**	0.9130	0.9480	**0.93**

Table 2: Submission results on the SMM4H 2021 Task 4 validation and test datasets.

RoBERTa Transformers	Fine-Tuning		Validation Set			Test Set		
	LM	Classifier	Precision	Recall	F1-score	Precision	Recall	F1-score
	No	Yes	0.7407	0.8197	0.7782	0.6750	0.7890	0.73
	Yes	Yes	0.7907	0.8361	**0.8128**	0.7452	0.7597	**0.75**

Table 2: Submission results on the SMM4H 2021 Task 5 validation and test datasets.

were also proposed to address this multi-class classification task (Reddy, 2020).

Our participated SMM4H 2021 Task 5 is new binary classification task, which aims at distinguishing tweets that self-report potential cases of COVID-19 from those that do not. COVID-19 Twitter Monitor was presented to show interactive visualizations of the analysis results on tweets related to the COVID-19 pandemic (Cornelius et al., 2020). An iterative graph-based approach was proposed to detect COVID-19 emerging symptoms using context-based twitter embeddings (Santosh et al., 2020). A large twitter dataset of COVID-19 chatter was used to identify discourse around drug mentions (Tekumalla and Banda, 2020).

3 The NCUEE-NLP System

Figure 1 shows our NCUEE-NLP system architecture for the SMM4H 2021 shared tasks. Specially, our system is composed of two main parts: RoBERTa transformers and fine-tuning. RoBERTa (a Robust optimized BERT pretraining approach) (Liu et al., 2019) is a replication study of BERT pretraining (Devlin et al., 2018) that carefully measures the impact of key parameters and training data size. We observe that RoBERTa transformers have usually performed well for many SMM4H 2020 tweet classification tasks (Klein et al., 2020c). Hence, we explore the usage of RoBERTa transformers and fine-tune the downstream tasks.

For Task 4, we use training, validation, and test datasets provided by task organizers to fine-tune the language model to improve the embedding representation. Then, the tweets with class labels in the training dataset were used to fine-tune the classifier.

For Task 5, because COVID-19 related tweets are relatively rare for fine-tuning the language model, we use the original training, validation, and test datasets from the Task 5 along with those tweets from Task 6 involving a three-class classification of COVID-19 tweets containing symptoms. To fine-tune the classifier, we only use the Task 5 training set that contains tweets with corresponding labels.

4 Evaluation

The experimental datasets were mainly provided by task organizers (Arjun et al., 2021). For Task 4, we have a total of 5,514 tweets in the training set, including 2,484 positive tweets and 3,030 negative tweets. The validation set contains 973 tweets (438 positive and 535 negative). Finally, there are a total of 10,000 tweets in the test set.

For Task 5, we have 6,465 tweets (1,026 positive and 5,439 negative) in the training set. The validation set contains 716 tweets (122 positive and 594 negative). Finally, there are 10,000 tweets in the test set. We also have a total of 16,067 tweets from Task 6 for fine-tuning the language model.

All tweets were pre-processed to convert emojis into the corresponding codes defined by the unicode consortium. The pre-trained RoBERTa-Large model was downloaded from HuggingFace (Wolf et al., 2019). The hyper-parameters used for both tasks are as follows: training batch size 64, learning rate 4e-5, and maximum sequence length 128.

Tables 1 and 2 respectively summarize the results for Tasks 4 and 5. The evaluation metric is the F1-score of the positive class for both tasks. It's obvious that we have consistent results for both tasks, with a performance boost coming from fine-tuning the language model. Our best results for both tasks slightly exceeded than the respective median scores of all submissions by 0.005.

5 Conclusions

This study describes the NCUEE-NLP system participating in SMM4H 2021 Task 4 for adverse pregnancy outcome and Task 5 for potential COVID-19 cases, including system design, implementation and evaluation. For Task 4, our best F1-score of 0.93 exceeded the median score of 0.925. For Task 5, our best F1-score of 0.73 exceeded the median score of 0.725.

Acknowledgments

This study is partially supported by the Ministry of Science and Technology, under the grant MOST 108-2218-E-008-017-MY3 and MOST 108-2634-F-008-003- through Pervasive Artificial Intelligence Research (PAIR) Labs, Taiwan.

References

Ari Z. Klein, and Graciela Gonzalez-Hernandez. 2020a. An annotated data set for identifying women reporting adverse pregnancy outcomes on twitter. *Data in Brief*, 32(2020): 106249. https://doi.org/10.1016/j.dib.2020.106249

Ari Z. Klein, Haitao Cai, Davy Weissenbacher, Lisa D. Levine, Graciela Gonzalez-Hernandez. 2020b. A natural language processing pipeline to advance the use of twitter data for digital epidemiology of adverse pregnancy outcomes. *Journal of Biomedical Informatics: X*, 8(2020):100076. https://doi.org/10.1016/j.yjbinx.2020.100076

Ari Z. Klein, Ilseyar Alimova, Ivan Flores, Arjun Magge, Zulfat Miftahutdinov, Anne-Lyse Minard, Karen O'Connor, Abeed Sarker, Elena Tutubalina, Davy Weissenbacher, and Graciela Gonzalez-Hernandez. 2020c. Overview of the fifth social media mining for health applications (#SMM4H) shared tasks at COLING 2020. In *Proceedings of the Fifth Social Media Mining for Health Applications Workshop & Shared Task*. Association for Computational Linguistics, pages 27-36.

Arjun Magge, Ari Z. Klein, Ivan Flores, Ilsear Alimova, Mohammed Ali Al-garadi, Antonio Miranda-Escalada, Zulfat Miftahutdinova, Eulalia Farre-Maduell, Salvador Lima Lopez, Juan M. Banda, Karen O'Connor, Abeed Sarker, Elena Tutubalina, Martin Krallinger, Davy Weissenbacher, and Graciela Gonzalez-Hernandez. 2021. Overview of the sixth social media mining for health application (#SMM4H) shared tasks at NAACL 2021. *Proceedings of the Sixth Social Media Mining for Health Applications Workshop & Shared Task*. Association for Computational Linguistics.

George-Andrei Dima, Andrei-Marius Avram, and Dumitru-Clementin Cercel. 2020. Approaching SMM4H 2020 with ensembles of BERT flavours. In *Proceedings of the Fifth Social Media Mining for Health Applications Workshop & Shared Task*. Association for Computational Linguistics, pages 153-157.

Jacob Devlin, Ming-Wei Chang, Kenton Lee, and Kristina Toutanova. 2018. BERT: pre-training of deep bidirectional transformers for language understanding. *arXiv preprint*, https://arxiv.org/abs/1810.04805

Joseph Cornelius, Tilia Ellendorff, Lenz Furrer, and Fabio Rinaldi. 2020. COVID-19 twitter monitor: aggregating and visualizing COVID-19 related trends in social media. In *Proceedings of the Fifth Social Media Mining for Health Applications Workshop & Shared Task*. Association for Computational Linguistics, pages 1-10.

Parsa Bagherzadeh, and Sabine Bergler. 2020. CLaC at SMM4H 2020: birth defect mention detection. In *Proceedings of the Fifth Social Media Mining for Health Applications Workshop & Shared Task*.

Association for Computational Linguistics, pages 168-170.

Ramya Tekumalla, and Juan M Banda. 2020. Characterizing drug mentions in COVID-19 twitter chatter. In *Proceedings of the 1st Workshop on NLP for COVID-19 (Part 2)*. Association for Computational Linguistics. http://dx.doi.org/10.18653/v1/2020.nlpcovid19-2.25

Roshan Santosh, H. Schwartz, Johannes Eichstaedt, Lyle Ungar, and Sharath Chandra Guntuku. 2020. Detecting emerging symptoms of COVID-19 using context-based twitter embeddings. In *Proceedings of the 1st Workshop on NLP for COVID-19 (Part 2)*. Association for Computational Linguistics. http://dx.doi.org/10.18653/v1/2020.nlpcovid19-2.35

Saichethan Miriyala Reddy. 2020. Detecting tweets reporting birth defect pregnancy outcome using two-view CNN RNN based architecture. In *Proceedings of the Fifth Social Media Mining for Health Applications Workshop & Shared Task*. Association for Computational Linguistics, pages 125-127.

Thomas Wolf, Lysandre Debut, Victor Sanh, Julien Chaumond, Clement Delangue, Anthony Moi, Pierric Cistac, Tim Rault, Remi Louf, Morgan Funtowicz, Joe Davison, Sam Shleifer, Patrick von Platen, Clara Ma, Yacine Jernite, Julien Plu, Canwen Xu, Teven Le Scao, Sylvain Gugger, Mariama Drame, Quentin Lhoest, and Alex- ander M. Rush. 2019. HuggingFace's transformers: state-of-the-art natural language processing. *arXiv preprint*. https://arxiv.org/abs/1910.03771

Yang Bai, and Xiaobing Zhou. 2020. Automatic detecting for health-related twitter data with BioBERT. In *Proceedings of the Fifth Social Media Mining for Health Applications Workshop & Shared Task*. pages 63-69.

Yinhan Liu, Myle Ott, Naman Goyal, Jingfei Du, Mandar Joshi, Danqi Chen, Omer Levy, Mike Lewis, Luke Zettlemoyer, Veselin Stoyanov. 2019. RoBERTa: a robustly optimized BERT pretraining approach. *arXiv preprint*, https://arxiv.org/abs/1907.11692

NLP@NISER: Classification of COVID19 tweets containing symptoms

Deepak Kumar[1,*], **Nalin Kumar**[2,*], **Subhankar Mishra**[3]
School of Computer Sciences, NISER, Bhubaneswar- 752050
Homi Bhabha National Institute, Training School Complex, Anushakti Nagar, Mumbai-400094, India
{deepak.kumar[1], nalin.kumar[2], smishra[3]}@niser.ac.in

Abstract

In this paper, we describe our approaches for task six of Social Media Mining for Health Applications (SMM4H) shared task in 2021. The task is to classify twitter tweets containing COVID-19 symptoms in three classes (self-reports, non-personal reports & literature/news mentions). We implemented BERT and XLNet for this text classification task. Best result was achieved by XLNet approach, which is F1 score 0.94, precision 0.9448 and recall 0.94448. This is slightly better than the average score, i.e. F1 score 0.93, precision 0.93235 and recall 0.93235.

1 Introduction

In the beginning of the COVID-19 pandemic and even now, with variety of strains, caused great deal of information deficiency about symptoms as reported by people affected by it. As this disease is highly contagious and rapidly changing, one of the best sources for the live information is on the social media. There can be multiple sources of symptoms information on social media such as news/scientific articles (facts), other people's account (second or third person statements) and self report (first person statements). In this paper we will discuss our approach as a team participating in SMM4H (Magge et al., 2021) shared task 6 related to the classification of such information from social media platform like twitter. We will be looking at using pre-trained NLU models like BERT (Devlin et al., 2018) and XLNet (Yang et al., 2019) for this task.

2 Task and Data Description

2.1 Task

The task 6 of SMM4H is to classify twitter tweet dataset containing COVID-19 symptoms into three classes:

- self-reports: mentioning ones own experience of COVID-19
- non-personal reports: mentioning other peoples account of COVID-19 experience
- literature/news mentions: mentioning scientific or news articles telling about COVID-19 symptoms

2.2 Data Description

The training dataset contained $9,000$ labeled tweets, validation dataset contained $5,76$ labeled tweets and test dataset contained $6,500$ unlabeled tweets.

3 Methodology

3.1 Pre-processing

- **Data Cleaning** : For cleaning we removed, part of sentences starting from "@" to first white space, links, numbers, "#" and extra white space.
- Further in pre-processing, we convert data into machine readable form. We change the labels into three discrete integers and tweet text text into tokens using tokenizer as mentioned in Model section.
- Then we truncate the tweet text to reduce the amount of padding. For BERT we truncate before applying tokenizer and making sentence length 65, while for XLNet we truncate after applying tokenizer to make sequence length 150.

3.2 Model

We explore both autoencoding as well as autoregressive models for the classification task from which BERT and XLNet are picked respectively. For our experiments, we use the pre-trained models provided by Huggingface (Wolf et al., 2020). For both of the models we are using cased versions which differentiates between upper and lower case.

*Equal contribution. Deepak implemented BERT while Nalin performed experiments on XLNet

Proceedings of the Sixth Social Media Mining for Health Workshop 2021, pages 102–104

This is needed as upper case letters are important for identifying nouns and pronouns which in turn are important to identify first, second and third person in a sentence.

3.2.1 BERT

The first system we use for the task is BERT (Devlin et al., 2018). BERT, which stands for Bidirectional Encoder Representations from Transformers, learns by predicting the randomly masked token during pre-training using both left and right context of the masked word token. It also has the second objective of predicting if the given two sentences are consecutive. We use both base and large cased versions of this model for our experiments. The base version of BERT has 12 encoder layers, 768 hidden layer dimension and 12 attention heads with 109M parameters, while the large one has 24 encoder layers, 1024 hidden layer dimension and 16 attention heads with 335M parameters. We use their respective tokenizers.

3.2.2 XLNet

We use XLNet as our second system. XLNet (Yang et al., 2019) is an auto-regressive model, which, unlike BERT, uses autoregressive formulation to learn the bidirectional contexts. The word token output is calculated by taking into account the permutation of all word tokens in the sentence, in contrast to the traditional approaches, which used just left or right of the target token. We experiment with both base and large versions of this model. The base version has 12 layers, 768 hidden layer dimension and 12 attention heads with number of model parameters to be $110M$, whereas the large one has 24 layers, 1024 hidden layer dimension and 16 attention heads having $340M$ parameters. We use their respective tokenizers.

4 Experiments

We perform several experiments on cleaned and uncleaned data. We explore both base and large versions of BERT and XLNet along with different training methods, fine-tuning and retraining the whole model, scores of which are mentioned in Table 1. For both the NLU models, we use appropriate classification layer. For all these experiments, loss function is cross entropy loss and optimizer is adamW with learning rate $2e-5$. We find BERT large version retrained on uncleaned data perform the best. For the XLNet version, we find that the large version of the XLNet tokenizer with base version of the model works the best among all. We get the best results on retraining the model over the uncleaned data. The best systems' scores on the test data for the shared task are given in Table 2. Our code is shared on Github [*].

Approach	BERT		XLNet	
	Base	Large	Base	Large
C_Retrain	0.976	0.978	0.974	0.968
C_Finetune	0.666	0.729	0.844	0.784
U_Retrain	0.98	0.998	0.984	0.984
U_Finetune	0.76	0.734	0.924	0.878

Table 1: All results are F1 scores of models trained over training set and calculated over validation set. All models are run for 6 epochs with batch size 16.

Model	Precision	Recall	F1 Score
XLNet	0.9448	0.94448	0.94
BERT	0.926	0.926	0.93
Median	0.93235	0.93235	0.93

Table 2: Performance of the best version of each model on the test set. Models are run for 6 epochs with batch size 16.

5 Results and Conclusion

We can have the following observations from Table 1 and 2:

- In comparison to training on cleaned data, the uncleaned versions show better results. We suspect that the information removed while data cleaning (such as "@", links, etc) are significant for predictions.
- We also observe that the retraining method performs significantly better than the fine-tuned one.
- BERT, the large version performs better than the base one, whereas in XLNet, the base version has better scores than the large one.
- Finally, the BERT performs better on validation set while the XLNet has better performance on the test set.

Between the given two systems, the XLNet performs the best with results shown in Table 2. The system performs slightly better than the median of all submissions made for the task. In the future work, one can look for new approach towards

[*] https://github.com/smlab-niser/2021smm4h

cleaning of tweets, as traditional way of cleaning tweets tend to decrease the F1-score (as shown by our results).

References

Jacob Devlin, Ming-Wei Chang, Kenton Lee, and Kristina Toutanova. 2018. Bert: Pre-training of deep bidirectional transformers for language understanding. *arXiv preprint arXiv:1810.04805.*

Arjun Magge, Ari Klein, Ivan Flores, Ilseyar Alimova, Mohammed Ali Al-garadi, Antonio Miranda-Escalada, Zulfat Miftahutdinov, Eulàlia Farré-Maduell, Salvador Lima López, Juan M Banda, Karen O'Connor, Abeed Sarker, Elena Tutubalina, Martin Krallinger, Davy Weissenbacher, and Graciela Gonzalez-Hernandez. 2021. Overview of the sixth social media mining for health applications (# smm4h) shared tasks at naacl 2021. In *Proceedings of the Sixth Social Media Mining for Health Applications Workshop & Shared Task.*

Thomas Wolf, Lysandre Debut, Victor Sanh, Julien Chaumond, Clement Delangue, Anthony Moi, Pierric Cistac, Tim Rault, Rémi Louf, Morgan Funtowicz, Joe Davison, Sam Shleifer, Patrick von Platen, Clara Ma, Yacine Jernite, Julien Plu, Canwen Xu, Teven Le Scao, Sylvain Gugger, Mariama Drame, Quentin Lhoest, and Alexander M. Rush. 2020. Transformers: State-of-the-art natural language processing. In *Proceedings of the 2020 Conference on Empirical Methods in Natural Language Processing: System Demonstrations*, pages 38–45, Online. Association for Computational Linguistics.

Zhilin Yang, Zihang Dai, Yiming Yang, Jaime Carbonell, Ruslan Salakhutdinov, and Quoc V Le. 2019. Xlnet: Generalized autoregressive pretraining for language understanding. *arXiv preprint arXiv:1906.08237.*

Identification of profession & occupation in Health-related Social Media using tweets in Spanish

Victoria Pachón Álvarez
Escuela Técnica Superior de Ingeniería
Universidad de Huelva (Spain)
vpachon@uhu.es

Jacinto Mata Vázquez
Escuela Técnica Superior de Ingeniería
Universidad de Huelva (Spain)
mata@uhu.es

Juan Luis Domínguez Olmedo
Escuela Técnica Superior de Ingeniería
Universidad de Huelva (Spain)
juan.dominguez@dti.uhu.es

Abstract

In this paper we present our approach and system description on Task 7a in ProfNer-ST: Identification of profession & occupation in Health related Social Media. Our main contribution is to show the effectiveness of using BETO-Spanish BERT for classification tasks in Spanish. In our experiments we compared several architectures based on transformers with others based on classical machine learning algorithms. With this approach, we achieved an F1-score of 0.92 for the positive class in the evaluation process.

1. Introduction

The battle against COVID-19 is present in practically every country in the world. Confinement, curfews and restrictions on the movement of personnel and cargo, are part of the strategy to stop the transmission of the virus. Some workers are at the forefront of the battle against the COVID-19 pandemic, and they are more exposed to the virus and also more likely to suffer from mental health problems because of the stress caused by the pandemic. The detection of vulnerable occupations is essential to prepare preventive measures. In ProfNer-ST: Task 7, Track A (Tweet binary classification) (Miranda-Escalada et al. 2021) participants must determine whether a tweet contains a mention of occupation, or not.

Despite Spanish being the 4th most spoken language, finding resources to train and evaluate hypothesized that automatically translating a text from Spanish to English to use and model based on this language would not be as good as working straight away with a model pre trained with a Spanish corpus. The idea behind all our experiments was to compare models pre-trained in Spanish with models pretrained in English and using automatic translations. In this context of work we have been heavily using BETO-Spanish BERT (Cañete et al. 2020) , BERT-Multilingual (Devlin et al. 2018) and RuPERTa: the Spanish RoBERTa (GitHub - mrm8488/RuPERTa-base: Spanish RoBERTa), and we compared them with the results obtained by BERT (Devlin et al. 2018) using the official translation of the given datasets in English.

2. Data Description and preprocessing

The corpus provided to perform Task 7a (classification) is described in (Magge et al. 2021). Since tweets have a very specific language, for this task we have not performed a very exhaustive data preprocessing. The only text processing performed was to convert all characters to lowercase. In order to carry out different experiments to evaluate the performance of our systems, we have used 3 files:

- **Original**. The original text of the tweets was preserved.
- **URLs_removed.** The URLs of the tweets were removed.
- **Hashtags_URLs_removed.** Both URLs and hashtags were removed from the tweets.

Proceedings of the Sixth Social Media Mining for Health Workshop 2021, pages 105–107
June 10, 2021.

3. Methods

Our methodology is based on working directly with texts in Spanish and applying multilingual or Spanish pre-trained models, instead of translated and using pre-trained English models. The system consists of fine-tuning a BETO model for classification tasks. Before starting to study the performance of our systems, we design a baseline and use its results as a starting point to improve our approaches. We trained a Bidirectional Long-Short Term Memory RNN with one dense layer and the *skipgram uncased* Spanish COVID-19 Twitter Embeddings (Miranda-Escalada et al. 2021b) for the word embedding layer. BERT variants gave good results in #SMM4H 2020 (Klein et al. 2020) so we decided to focus on them to develop our proposal. We have carried out several experiments to compare the results of BETO with multilingual Bert (mBert) and RuPERTa as well as English pretrained BERT (cased and uncased). For all the experiments, we used the training and test dataset supplied by the organizers. The training dataset was split up in two parts to get a validation dataset (30%). We used a batch size of 32 instances, and we trained with 4 epochs and max length of 256. In our experiments with the BERT English pretrained model we have used the translation to English provided by the organizers. In all experiments with uncased models, we have transformed each tweet to lowercased. BERT (Bidirectional Encoder Representations for Transformers) also offers a multilingual model (mBERT) pretrained on concatenated Wikipedia data for languages without any cross-lingual alignment. BETO (Spanish Pre-Trained BERT Model and Evaluation Data) is a model similar in size to a BERT-Base model with 12 self-attention layers, 16 attention-heads each (Vaswani et al. 2017) and 1024 as hidden size. The total size of the corpora gathered was comparable with the corpora used in the original BERT. RuPERTa-base (uncased) is a RoBERTa model trained on an uncased version of big Spanish corpus and its architecture is the same as Roberta-base (Liu et al. 2019).

4. Experiments and Results

We fine-tuned mBert, BETO, RuPERTa and BERT with the training dataset provided by the organizers and we test the model obtained using the test dataset described before. The measure was F1-score for the positive class, according to the one used for the ranking of the systems in the competition. Table 1 shows a summarization of the experimental results obtained. BETO obtained the best results for all the measures. We fine-tuned BETO-cased using all the tweet from the training and test datasets with 5 epoch and we made predictions on the unseen evaluation examples as our first and only submission. We achieved an F1-score of 0.92 in the evaluation process.

	Original Dataset			URL_removed Dataset			Hashtags_URLs_removed Dataset		
	F1-score (class 1)	macro-F1	AUC	F1-score (class 1)	macro-F1	AUC	F1-score (class 1)	macro-F1	AUC
baseline	0.77	0.86	0.830	0.75	0.84	0.816	0.78	0.86	0.844
mBert -uncased-	0.88	0.92	0.915	0.88	0.92	0.918	0.87	0.92	0.910
mBert -cased-	0.88	0.92	0.919	0.88	0.92	0.915	0.87	0.91	0.911
BETO -uncased-	0.91	0.94	0.934	0.90	0.93	0.930	0.89	0.93	0.918
BETO -cased- (our propousal)	**0.91**	**0.94**	**0.939**	0.90	0.94	0.936	0.89	0.93	0.923
RuPERTa-spanish	0.75	0.83	0.834	0.76	0.84	0.844	0.76	0.84	0.848
BERT -uncased-	0.88	0.92	0.915	0.88	0.92	0.919	0.88	0.92	0.909
BERT -cascd-	0.88	0.92	0.916	0.87	0.91	0.904	0.87	0.91	0.904

Table 1. Results on test dataset

5. Conclusions

In this paper we present our approach and system description on Task 7a in ProfNer-ST: Identification of profession & occupation in Health related Social Media. The main idea was checking the use of models trained with a Spanish corpus. Our model was based on fine tuning a pretrained model in Spanish: BETO for classification tasks. In our experiments we also tested and compared several architectures based on transformers with others based on classical machine learning algorithms. In the future we want to keep testing BETO in other contests. With this approach, we achieved an F1-score of 0.92 in the evaluation process for class "1". In this way, we proved the accuracy and usability of pretrained models with a Spanish Corpus.

References

José Cañete, Gabriel Chaperon, Rodrigo Fuentes, Jou-Hui Ho, Hojin Kang, and Jorge Pérez. 2020. *Spanish Pre-Trained BERT Model and Evaluation Data.* PML4DC at ICLR 2020

Jacob Devlin, Ming-Wei Chang, Kenton Lee, and Kristina Toutanova. 2018. *BERT: Pre-Training of Deep Bidirectional Transformers for Language Understanding*. Proc. of the Conference of the North American Chapter of the Association for Computational Linguistics: Human Language Technologies. DOI: 10.18653/v1/N19-1423

Ari Klein, Ilseyar Alimova, Ivan Flores, et al. 2020. *Overview of the Fifth Social Media Mining for Health Applications (\#SMM4H) Shared Tasks at COLING 2020.* Proc. of the Fifth Social Media Mining for Health Applications Workshop & Shared Task

Yinhan Liu, Myle Ott, Naman Goyal, Jingfei Du, et al. 2019. *RoBERTa: A Robustly Optimized BERT Pretraining Approach.* arxiv.org/abs/1907.11692

Arjun Magge, Ari Klein, Ivan Flores, Ilseyar Alimova, et al. 2021. *Overview of the Sixth Social Media Mining for Health Applications (\#SMM4H) Shared Tasks at NAACL 2021*

Antonio Miranda-Escalada, Eulália Farré-Maduell, Salvador Lima, Vicent Briva-Iglesias, et al. 2021. *The ProfNER Shared Task on Automatic Recognition of Professions and Occupation Mentions in Social Media: Systems, Evaluation, Guidelines, Embeddings and Corpora*

Antonio Miranda-Escalada, Marvin Aguero, and Martin Krallinger,. 2021. *Spanish COVID-19 Twitter Embeddings in FastText.* DOI: http://doi.org/10.5281/zenodo.4449930

Jörg Tiedemann. 2012. *Parallel Data, Tools and Interfaces in OPUS.* Proc. of the Eighth International Conference on Language Resources and Evaluation (LREC'12)

Ashish Vaswani, Noam Shazeer, Niki Parmar, Jakob Uszkoreit, et al. 2017. *Attention is all You Need.* arXiv:1706.03762v5

Lasige-BioTM at ProfNER: BiLSTM-CRF and contextual Spanish embeddings for Named Entity Recognition and Tweet Binary Classification

Pedro Ruas and **Vitor D. T. Andrade** and **Francisco M. Couto**
LASIGE, Faculdade de Ciências da Universidade de Lisboa
1749-016 Lisboa, Portugal
psruas@fc.ul.pt, fc49005@alunos.fc.ul.pt, fcouto@di.fc.ul.pt

Abstract

The paper describes the participation of the Lasige-BioTM team at sub-tracks A and B of ProfNER, which was based on: i) a BiLSTM-CRF model that leverages contextual and classical word embeddings to recognize and classify the mentions, and ii) on a rule-based module to classify tweets. In the Evaluation phase, our model achieved a F1-score of 0.917 (0,031 more than the median) in sub-track A and a F1-score of 0.727 (0,034 less than the median) in sub-track B.

1 Introduction

The track "ProfNER-ST: Identification of professions & occupations in Health-related Social Media" (Miranda-Escalada et al., 2021b) occurred in the context of the "Social Media Mining for Health Applications (#SMM4H) Shared Task 2021" (Magge et al., 2021), and included two different sub-tracks that focused on Spanish Twitter data:

- Track A – Tweet binary classification: to determine if a given tweet has a mention of occupation or not.
- Track B – Named Entity Recognition (NER) offset detection and classification: to recognise the span of mentions of occupations and classify them in the respective category.

This paper describes the participation of the Lasige-BioTM team in the aforementioned sub-tracks. We applied 8 different models NER models (4 supervised models based on BiLSTM-CRF architecture, 3 rule-based models) to predict entities for sub-track B and explored the impact of performing data augmentation in the training set. For sub-track A, we developed a rule-based model for tweet classification that was based on the NER output for sub-track B.

1.1 Related Work

According to Goyal et al. (2018), NER approaches can be divided in two categories: rule-based and machine learning-based, being the latter further subdivided into supervised, semi-supervised, unsupervised; other approaches combine aspects from the two categories and are thus designated by hybrid. The models with an architecture consisting of a bidirectional Long Short-Term Memory (BiLSTM) network and a Conditional Random Field (CRF) decoding layer are among the state-of-the-art approaches for the NER task. (Huang et al., 2015). For a comprehensive overview of the existing NER approaches please refer to Goyal et al. (2018) and, specifically for the biomedical domain, to Lamurias and Couto (2019).

2 Methodology

2.1 Corpus description

The ProfNER corpus (Miranda-Escalada et al., 2020) contains 10,000 health-related tweets in Spanish that were annotated by linguist experts with entities relative to professions, employment statuses, and other work-related activities and includes four categories: "PROFESION", "SITUACION_LABORAL", "ACTIVIDAD", and "FIGURATIVA". For sub-track A, a given tweet was assigned the label "1" if it included at least one entity belonging to any category, but for sub-track B only entities belonging to categories "PROFESION" and "SITUACION_LABORAL" were considered for evaluation.

2.2 Pre-processing

We performed data augmentation on the training set of the corpus using the Python library `nlpaug` (Ma, 2019). For example, considering the mentioned entity "médico" present in the training set, data augmentation consisted of substituting a random character by a keyboard character (i.e. replac-

Proceedings of the Sixth Social Media Mining for Health Workshop 2021, pages 108–111
June 10, 2021. ©2021 Association for Computational Linguistics

ing the character by a neighbour character in the keyboard in order to simulate a typing error character, since Twitter data is usually noisy: "médico" $\rightarrow$ "médLco"), by a random distance character ("médico" $\rightarrow$ "médicB"), and by a synonym (i.e. replacing the character by a synonym in the Spanish WordNet: "médico" $\rightarrow$ "dr."). The output of this step consisted of three additional training files besides the original training file, each one associated with the result of a type of augmentation.

2.3 MER

The first approach was based on `MER` (Couto and Lamurias, 2018), a minimal NER tagger that recognizes entities and the respective span in text according to a given lexicon. It is based on the text processing command-line tools `grep` and `awk`, and on an inverted recognition technique that uses the words in input text as patterns to match the lexicon words. Several lexicons were created and processed including: 1) mentions in "PROFESION" category in training set and its WordNet synonyms, 2) mentions in "PROFESION" category in training set and its WordNet synonyms, jointly with entities present in the Occupations gazetteer provided by the organisation (Asensio et al., 2021), 3) mentions in "SITUACION_LABORAL" category in training set and its WordNet synonyms, 4) entities in "ACTIVIDAD" category in train set and its WordNet synonyms, 5) entities in "FIGURATIVA" category in train set and its WordNet synonyms. The first model ("MER 1") included the lexicons 1, 3, 4, and 5, the second model ("MER 2") included the lexicons 2, 3, 4, and 5, the third model ("MER 3") was similar to the first one but the mention "sin" was filtered out. During Practice phase, we built the lexicons from the training set and used the validation set as the test set. For sub-task A, we developed a rule-based module to classify each tweet with the label "1" if at least one mention was recognized in the respective text, and with label "0" otherwise.

2.4 BiLSTM-CRF

To implement the second approach, we resorted to the `FLAIR` framework (Akbik et al., 2019), and created an object of the class `SequenceTagger`, which instantiates a NER model with an architecture consisting of a BiLSTM network and a CRF decoding layer. LSTM are recurrent neural networks (RNNs), which include an input layer x representing features at time t, one or more hidden layers h, and an output layer y, which in the case of the NER task, represents a probability distribution over labels or tags at time t. A CRF network focus on the sentence level and also uses past and future tags/labels to predict the current one. The combination of a BiLSTM network with a CRF network has shown performance improvements over alternative architectures (Huang et al., 2015).

In the NER task, text needs to be tokenized and vectorized before being inputed to the neural network, which can be done leveraging pre-trained embeddings. FastText embeddings (Bojanowski et al., 2017) are an improvement over classic word embeddings, more concretely the skipgram model, by capturing sub-word information. `FLAIR` embeddings (Akbik et al., 2018) are contextual string embeddings that capture syntactic-semantic word features. We have explored the integration of different types of embeddings in the BiLSTM-CRF model through the `StackedEmbeddings` class:

- "Base" : `FLAIR` embeddings ("es-forward" and "es-backward") trained on Spanish Wikipedia (Akbik et al., 2018) + Spanish FastText embeddings
- "Twitter" : FastText Spanish COVID-19 Twitter Embeddings, provided by the organization (Miranda-Escalada et al., 2021a) (uncased version of the cbow model).
- "Medium" : `FLAIR` embeddings ("es-forward" and "es-backward") + Spanish FastText embeddings + FastText Spanish COVID-19 Twitter Embeddings

For the sub-track A, we applied a similar rule-based module as described in Section 2.3. If a model recognizes at least one entity in a given tweet in the context of sub-track B, the module assigns the label "1" to the respective tweet. If no entity is recognized in a given tweet, this receives the label "0". All the tweet IDs and respective label are then outputted in the predictions file for sub-track A.

2.4.1 Training

During Practice phase, we trained the models "Base" and "Twitter" on the original training file ("Base" and "Twitter"), and additionally, on the three files that resulted from the data augmentation step ("Base-aug" and "Twitter-aug"). During Evaluation phase, we merged the training and validation annotations, resulting in a file composed by 14,674 sentences for training and 1,630 sentences for validation. The training parameters were set to: `hidden size` = 256, `Mini batch size` = 32, `Max epochs` = 55, `Patience` = 3.

3 Results and discussion

3.1 Practice phase

The performance of the referred models in the validation set for sub-tracks A and B are available in Table 1. The "Base" model trained on the original training file achieved the best performance in sub-tracks A and B: F1-scores (strict) of 0.908 and 0.716, respectively. Consequently, we selected this model for further training and application in the test set. The models trained on files resulting from data augmentation achieved lower performances compared with the respective versions trained exclusively on the original training file.

3.2 Evaluation phase

The results achieved by our model in the Evaluation phase and the median results for all competing teams are shown in Table 2. In sub-track A, our model achieved a F1-score of 0.917 (0.031 more than the median) and in sub-track our model achieved a F1-score of 0.727 (0.034 less than the median).

3.3 Error analysis

The model "Base", that uses contextual embeddings trained on a general corpora, obtained higher performance when comparing to the model "Twitter", although this latter model uses Twitter-specific embeddings, more concretely, FastText embeddings that were trained on Twitter data. For instance, consider the following tweet of the validation set: *"Ya que están sesionando la importante pero NO prioritaria #LeyDeAmnistia,será que también vean la cuestión de #Economia y #SaludParaTodos? Digo!Recuerden que su prioridad somos los millones que estamos indefensos ante el #COVID-19 y sin trabajo @MorenaSenadores #LeyDeAmnistiaNo https://t.co/DCiuqiBjEs"*. The model "Twitter" recognizes the mention "@MorenaSenadores" and assigns the "PROFESION" category to it, whereas the model "Base" does not recognize any mention, since is been able to assume in this context that the mention do not correspond to a profession, but instead to a Twitter handle. There is a mention with the string "senadores" classified as "PROFESION" in a tweet of the training set, which maybe leads the model "Twitter" to assume that the words "@MorenaSenadores" must also correspond to a mention, since the string is similar.

4 Conclusion

During the Practice Phase, we explored different approaches to participate in sub-tracks A e B of ProfNER: data augmentation on training set, and application of MER and a BiLSTM-CRF model for NER and further tweet classification. For the Evaluation phase we applied the BiLSTM-CRF model on the test set of ProfNER corpus and achieved F1-scores of 0.917 (0,031 more than the median) and in sub-track our model achieved a F1-score of 0.727 (0,034 less than the median). The code to run the experiments is available in our GitHub page[1]. For future work, we intend to perform hyper-parameter optimisation for the BiLSTM-CRF model, such as learning rate, hidden size, and specially the number of training epochs, since we had limited available time to perform the training of the model. We will also explore the use of different contextualised embeddings, since the models using this type of embeddings seem to achieve better performance compared to those using classical word embeddings. Besides, to improve tweet classification we will explore the application of Named Entity Linking tools (Lamurias et al., 2019) to link the recognized entities in sub-track B to structured vocabularies that contain hierarchical relationships between concepts, such as MeSH or DBpedia. This way, it will be possible to know the ancestors for a given entity, which will provide the context to effectively determine if the entity is associated with an occupation or not.

Acknowledgements

This work was supported by FCT through funding of Deep Semantic Tagger (DeST) project (ref. PTDC/CCI-BIO/28685/2017) and LASIGE Research Unit (ref. UIDB/00408/2020 and ref. UIDP/00408/2020); and FCT through funding of PhD Scholarship, ref. 2020.05393.BD.

References

Alan Akbik, Tanja Bergmann, Duncan Blythe, Kashif Rasul, Stefan Schweter, and Roland Vollgraf. 2019. Flair: An easy-to-use framework for state-of-the-art nlp. In *NAACL 2019, 2019 Annual Conference of the North American Chapter of the Association for*

[1] `https://github.com/lasigeBioTM/LASIGE-participation-in-ProfNER`

	Sub-track 7A			Sub-track 7B					
Model	P	R	F1	P	R	F1	Rel-P	Rel-R	Rel-F1
MER 1	0.621	0.767	0.687	0.399	0.535	0.457	0.565	0.668	0.612
MER 2	0.498	0.839	0.625	0.290	0.578	0.386	0.418	0.721	0.529
MER 3	0.621	0.767	0.687	0.472	0.535	0.501	0.668	0.667	0.667
Base	**0.941**	0.876	**0.908**	**0.795**	**0.651**	**0.716**	**0.901**	**0.738**	**0.811**
Base-aug	0.848	0.830	0.839	0.705	0.616	0.657	0.826	0.721	0.770
Twitter	0.895	0.874	0.884	0.721	0.616	0.664	0.856	0.730	0.788
Twitter-aug	0.786	0.904	0.841	0.597	0.611	0.604	0.737	0.755	0.746
Medium-aug	0.780	0.887	0.830	0.618	0.601	0.609	0.753	0.733	0.743

Table 1: Practice results for sub-track 7A (left) and sub-track 7B (right). P, R, and F1 refer to precision, recall, and F1-score (strict), respectively and Rel-P, Rel-R, and Rel-F1 refer to relaxed precision, relaxed recall, and relaxed F1-score, respectively

	Sub-track 7A			Sub-track 7B		
Model	P	R	F1	P	R	F1
Lasige-BioTM	**0.951**	**0.886**	**0.917**	0.814	0.657	0.727
Median	0.919	0.855	0.886	**0.842**	**0.727**	**0.761**

Table 2: Evaluation phase results for sub-tracks 7A and 7B. P, R, F1 refer to precision, recall, and F1-score (strict), respectively.

Computational Linguistics (Demonstrations), pages 54–59.

Alan Akbik, Duncan Blythe, and Roland Vollgraf. 2018. Contextual string embeddings for sequence labeling. In *COLING 2018, 27th International Conference on Computational Linguistics*, pages 1638–1649.

Alejandro Asensio, Antonio Miranda-Escalada, Marvin Aguero, and Martin Krallinger. 2021. Occupations gazetteer - ProfNER & MEDDOPROF - occupations, professions and working status terms with their associated codes. Funded by the Plan de Impulso de las Tecnologías del Lenguaje (Plan TL).

Piotr Bojanowski, Edouard Grave, Armand Joulin, and Tomas Mikolov. 2017. Enriching word vectors with subword information.

Francisco M. Couto and Andre Lamurias. 2018. MER: a shell script and annotation server for minimal named entity recognition and linking. *Journal of Cheminformatics*, 10(1):58.

Archana Goyal, Vishal Gupta, and Manish Kumar. 2018. Recent Named Entity Recognition and Classification techniques: A systematic review. *Computer Science Review*, 29:21–43.

Zhiheng Huang, Wei Xu, and Kai Yu. 2015. Bidirectional lstm-crf models for sequence tagging.

Andre Lamurias and Francisco M Couto. 2019. Text Mining for Bioinformatics Using Biomedical Literature. In K. and Ranganathan, S., Gribskov, M., Nakai and C Schoonbach, editors, *Encyclopedia of Bioinformatics and Computational Biology, vol. 1*, January, pages pp. 602–61. Oxford: Elsevier.

Andre Lamurias, Pedro Ruas, and Francisco M. Couto. 2019. PPR-SSM: Personalized PageRank and semantic similarity measures for entity linking. *BMC Bioinformatics*, 20(1):1–12.

Edward Ma. 2019. Nlp augmentation. https://github.com/makcedward/nlpaug.

Arjun Magge, Ari Z. Klein, Ivan Flores, Ilseyar Alimova, Mohammed Ali Al-garadi, Antonio Miranda-Escalada, Zulfat Miftahutdinov, Eulalia Farre, Salvador Lima, Juan Banda, Karen O'Connor, Abeed Sarker, Elena Tutubalina, Martin Krallinger, Davy Weissenbacher, and Graciela Gonzalez-Hernandez. 2021. Overview of the sixth social media mining for health applications (#smm4h) shared tasks at naacl 2021.

Antonio Miranda-Escalada, Marvin Aguero, and Martin Krallinger. 2021a. Spanish covid-19 twitter embeddings in fasttext. Funded by the Plan de Impulso de las Tecnologías del Lenguaje (Plan TL).

Antonio Miranda-Escalada, Vicent Briva-Iglesias, Eulàlia Farré, Salvador Lima López, Marvin Aguero, and Martin Krallinger. 2020. ProfNER corpus: gold standard annotations for profession detection in Spanish COVID-19 tweets. Funded by the Plan de Impulso de las Tecnologías del Lenguaje (Plan TL).

Antonio Miranda-Escalada, Eulàlia Farré-Maduell, Salvador Lima López, Luis Gascó-Sánchez, Vicent Briva-Iglesias, Marvin Agüero-Torales, and Martin Krallinger. 2021b. The profner shared task on automatic recognition of occupation mentions in social media: systems, evaluation, guidelines, embeddings and corpora. In *Proceedings of the Sixth Social Media Mining for Health Applications Workshop & Shared Task*.

Phoenix@SMM4H Task-8: Adversities Make Ordinary Models Do Extraordinary Things

Adarsh Kumar*, **Ojasv Kamal*** and **Susmita Mazumdar***
Indian Institute of Technology Kharagpur
{adarshkumar712, kamalojasv47, susmita10}@iitkgp.ac.in

Abstract

In this paper, we describe our system entry for Shared Task 8 at SMM4H-2021 , which is on automatic classification of self-reported breast cancer posts on Twitter. In our system, we use a transformer-based language model fine-tuning approach to automatically identify tweets in the self-reports category. Furthermore, we involve a **Gradient-based Adversarial fine-tuning** to improve the overall model's robustness. Our system achieved an F1-score of **0.8625** on the development set and **0.8501** on the test set in Shared Task-8 of SMM4H-2021.

1 Introduction

With increased cases of discontinuation of Breast Cancer Treatment, which often leads to cancer recurrence, there is a need to explore complementary sources of information for patient-centered-outcomes(PCOs) associated with breast cancer treatments. Social media is a promising resource but extracting true PCOs from it first requires the accurate detection of self-reported breast cancer patients. (Al-Garadi et al., 2020) presented an NLP architecture along with a dataset for automatically categorising self-reported breast cancer posts. Their dataset was released as Shared Task-8 in SMM4H 2021.

In this paper, we describe our system to automatically distinguish self-reports of breast cancer from non-relevant tweets, which we used in the final submission for the Shared Task-8 of SMM4H-2021 (our best submission).

2 Methodology

2.1 Task and Dataset Overview

The task 8 of SMM4H consists of automatic classification of tweets into self-reports of breast cancer or non-relevant categories. The dataset comprises tweets, each associated with a label. The label indicates whether the corresponding tweet is a self-report of breast cancer or not (1 for yes, 0 for no). The training set is an unbalanced dataset of 3815 labelled tweets, around 26% of which are self-reports of breast cancer. The test set comprises 1204 unlabelled tweets, and our objective is to categorise them as self-reports or non-relevant posts.

2.2 Data Preprocessing

Before feeding into the model for training, we remove the tweet ID, username, URLs and all Non-ASCII characters associated with each tweet. Furthermore, we replaced emoticons from tweets using Ekhprasis Package (Baziotis et al., 2017), with their respective dict labels present in ekphrasis.dicts.emoticons.

2.3 Our Proposed Approach

Fig 1 illustrates our proposed approach used for final submission in the Shared Task at SMM4H. It is an amalgamation of two approaches: Domain-Specific Pre-trained model fine-tuning for binary classification and Adversarial fine-tuning for model's robustness. Below, we define each module in detail.

Domain Specific Pre-Trained Model Fine-tuning: In order to classify tweets as self-reports or not, we try to leverage the information from large pre-trained models like BERT. We further try to improve the performance by using models pre-trained on Medical Dataset to leverage domain-specific information. We performed our experiments with BERT and BlueBERT (Peng et al., 2019) models from huggingface(Wolf et al., 2019) library. While fine-tuning, we use the output from the first token of the transformer model as contextualized embedding, which is then fed into a single feed-forward classifier layer, trained using Binary Cross-entropy Loss which can be mathematically formulated as:

* Equal Contribution

Proceedings of the Sixth Social Media Mining for Health Workshop 2021, pages 112–114
June 10, 2021.

$$L(\hat{y}, y) = -\sum_{j=1}^{c} y_j log\hat{y_j} + (1 - y_j)log(1 - \hat{y_j})$$

where, c is the total number of training examples

Gradient-based Adversarial Fine-tuning: Though fine-tuned Language Models perform well on downstream tasks like Text-classification, these models are often vulnerable to Adversarial attack (Li et al., 2020). Adversarial Fine-tuning (Goodfellow et al., 2015) has proved very efficient in improved generalization by Neural Network based models in Computer Vision Tasks.(Vernikos et al., 2020) and (Chen et al., 2021) showcase a gradient-based adversarial fine-tuning approach in the text-domain. In our system, we employ a similar technique to improve the robustness of our model. The key idea is to modify the training objective by applying small gradient-based perturbations to input text that maximize the adversarial loss. These perturbations (r1, r2, ... in Fig 1) can be easily computed using backpropagation in neural networks. The loss function we used in adversarial fine-tuning can be formulated as:

$$L = -log\, p(y|x + r_{adv})$$

where

$$r_{adv} = -\epsilon\frac{g}{||g||} \quad \text{where} \quad g = \nabla_x log(y|x;\theta)$$

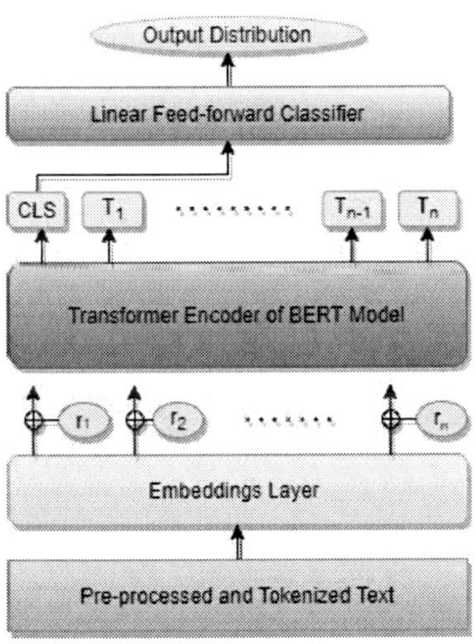

Figure 1: System Architecture

3 Result and Discussion

Table 1 shows the performance of different pre-trained models and approaches on Development Dataset, used in our experiment. As it is clear from Table 1, Blue BERT (Peng et al., 2019) fine-tuned using the Gradient-Based Adversarial Fine-tuning approach, outperforms other approaches and models on the development set. The results also suggest that adversarial fine-tuning, instead of normal fine-tuning, improves the performance of models, except for the BERT Large model. Two other important aspects to note are: improvement in the performance on using large models against the base models (which is expected given the increased number of parameters and model size) and the usefulness of Domain-specific Pre-trained model with Adversarial Fine-tuning. Table 2 shows the performance of our system entries in Shared Task - 8 on hold-on Test Dataset, which we selected after taking into consideration the above analysis.

Model	Dev F1 score
BERT base + FT	0.7826
BlueBERT base + FT	0.8025
BERT Large + FT**	0.8496
BlueBERT Large + FT	0.8205
BERT base + FT	0.8152
BlueBERT base + AFT	0.8289
BERT Large + AFT	0.8289
BlueBERT Large + AFT**	**0.86250**

Table 1: F1 score for Self Report Labelling on Development set (ϵ=1). **FT:** Fine-tuning and **AFT:** Adversarial Fine-tuning .Our final submission entries for Task 8 at SMM4H Shared Task is marked with **.

Model	F1	P	R
BERT+FT	0.8475	0.8754	**0.8214**
BlueBERT+AFT	**0.8508**	**0.8901**	0.8149

Table 2: Result on Hold-on test dataset on submission entries in SMM4H Shared Task-8. **F1:** F1 Score, **P**: Precision, **R**: Recall. Also, note both these submissions are with Large models, i.e. BERT-Large and BlueBERT Large models

4 Conclusion

In this paper, we described our approach of Adversarial fine-tuning on Domain-Specific Pre-Trained Model for classification of tweets as self-reports or not, which we used in our best submission at SMM4H-2021, Shared Task 8. Our ablation study demonstrates the usefulness of adversarial fine-tuning in improving the robustness of the model.

References

Mohammed Ali Al-Garadi, Yuan-Chi Yang, Sahithi Lakamana, Jie Lin, Sabrina Li, Angel Xie, Whitney Hogg-Bremer, Mylin Torres, Imon Banerjee, and Abeed Sarker. 2020. Automatic breast cancer cohort detection from social media for studying factors affecting patient centered outcomes. *medRxiv*.

Christos Baziotis, Nikos Pelekis, and Christos Doulkeridis. 2017. Datastories at semeval-2017 task 4: Deep lstm with attention for message-level and topic-based sentiment analysis. In *Proceedings of the 11th International Workshop on Semantic Evaluation (SemEval-2017)*, pages 747–754, Vancouver, Canada. Association for Computational Linguistics.

Ben Chen, Bin Chen, Dehong Gao, Qijin Chen, Chengfu Huo, Xiaonan Meng, Weijun Ren, and Yang Zhou. 2021. Transformer-based language model fine-tuning methods for covid-19 fake news detection.

Ian J. Goodfellow, Jonathon Shlens, and Christian Szegedy. 2015. Explaining and harnessing adversarial examples.

Linyang Li, Ruotian Ma, Qipeng Guo, Xiangyang Xue, and Xipeng Qiu. 2020. Bert-attack: Adversarial attack against bert using bert.

Yifan Peng, Shankai Yan, and Zhiyong Lu. 2019. Transfer learning in biomedical natural language processing: An evaluation of bert and elmo on ten benchmarking datasets. In *Proceedings of the 2019 Workshop on Biomedical Natural Language Processing (BioNLP 2019)*, pages 58–65.

Giorgos Vernikos, Katerina Margatina, Alexandra Chronopoulou, and Ion Androutsopoulos. 2020. Domain adversarial fine-tuning as an effective regularizer.

Thomas Wolf, Lysandre Debut, Victor Sanh, Julien Chaumond, Clement Delangue, Anthony Moi, Pierric Cistac, Tim Rault, Rémi Louf, Morgan Funtowicz, and Jamie Brew. 2019. Huggingface's transformers: State-of-the-art natural language processing. *CoRR*, abs/1910.03771.

A Supplemental Material

Links to the huggingface models used in the experiment:

- BERT base: `https://huggingface.co/bert-base-uncased`
- BERT Large: `https://huggingface.co/bert-large-uncased`
- BlueBERT Large: `https://huggingface.co/bionlp/bluebert_pubmed_uncased_L-24_H-1024_A-16`
- blueBERT base: `https://huggingface.co/bionlp/bluebert_pubmed_uncased_L-12_H-768_A-12`

UoB at ProfNER 2021: Data Augmentation for Classification Using Machine Translation

Frances Laureano De Leon,
Harish Tayyar Madabushi and **Mark Lee**

School of Computer Science
University of Birmingham
United Kingdom

fxl846@cs.bham.ac.uk
Harish@HarishTayyarMadabushi.com, M.G.Lee@bham.ac.uk

Abstract

This paper describes the participation of the UoB-NLP team in the ProfNER-ST shared subtask 7a. The task was aimed at detecting the mention of professions in social media text. Our team experimented with two methods of improving the performance of pre-trained models: Specifically, we experimented with data augmentation through translation and the merging of multiple language inputs to meet the objective of the task. While the best performing model on the test data consisted of mBERT fine-tuned on augmented data using back-translation, the improvement is minor possibly because multi-lingual pre-trained models such as mBERT already have access to the kind of information provided through back-translation and bilingual data.

1 Introduction and Motivation

The increase of user-generated content online has allowed researchers to extract information for studies on a variety of subjects, namely tracking infectious diseases and promoting public health (Wakamiya et al., 2018; Fine et al., 2020). Consequently the emergence of COVID-19 has resulted in a rapid increase of information related to the virus on social media platforms (Zhao et al., 2020). ProfNER, a task under Social Media Mining for Health Applications (SMM4H) workshop (Magge et al., 2021), requires the identification of occupations that might be particularly affected, either mentally or physically, by the exposure to COVID-19. The task organisers give participants tweets in Spanish and English. The English tweets were translated by means of a machine translation system. Of the tweets provided, 24% contain a mention of an occupation (Miranda-Escalada et al., 2021).

Classifiers are dependent on the size and quality of the training data (Wei and Zou, 2020), and are sensitive to class imbalance. We hypothesise that increasing the number of examples in the positive class using data augmentation techniques will successfully increase the performance of trained models. This work describes the training of four classifiers using pre-trained BERT models to detect the mention of occupations in tweets. In addition to training two baseline models, BERT-Base and mBERT, we train one model on augmented textual data, mBERT-Aug, and another model on bilingual data. We compare these models to each other and to fine-tuned pre-trained BERT models, described in Section 3.2. The small increase in F1 scores over the baselines, which is inconsistent across our validation and test experiments leads us to conclude that back-translation and bilingual data input are ineffective as methods of addressing class imbalance in pre-trained models, especially multi-lingual models (See Section 4). Our models were trained using the data provided by the task organisers for subtask 7a. Results are discussed in Section 4.

2 Related Work

Augmenting textual data is challenging because it can introduce label noise and must be done before training a model (Shleifer, 2019). Among techniques developed for text augmentation is synonym replacement, random insertion, swap and deletion, as presented by Wei and Zou (2020). Shleifer (2019) uses back-translation, to translate the data in a second language and then back to the source language. They train their model on a binary classification task in a setting where low amounts of labelled data are available. Work continues to be done in back-translation for classification, as there is little research otherwise (Shleifer, 2019). In this work, we use back-translation as a tool for augmenting the text data for the positive class. This work contributes to the field of generating synthetic

Proceedings of the Sixth Social Media Mining for Health Workshop 2021, pages 115–117
June 10, 2021.

data for text classification. Others have tried to add features to models to increase performance (Whang and Vosoughi, 2020; Lu et al., 2020), we attempt to bring together representations in different languages so as to maximise the information available to the models.

3 System Overview and Experimental Set-Up

This section describes our experimental design. The code, models, and hyper-parameters are available on our team's GitHub repository for the task [1].

3.1 Preprocessing

Punctuation, hashtags, twitter handles, emojis and URL's were all removed from the English and Spanish tweets. Tweets were tokenised using the Hugging Face Transformers library (Wolf et al., 2020).

3.2 Model Architecture

We trained four classifiers: mBERT-base, BERT-base, mBERT-Aug, and bilingual models. We utilised pre-trained mBERT-base and BERT-base to conduct our experiments (Devlin et al., 2019) using both the Spanish and English training data.

Our team fine-tuned mBERT and BERT-base to use as a baseline for our experiments. We fine-tuned both models with the 6,000 train tweets provided by the task organisers; mBERT was trained on Spanish tweets and BERT-base on English tweets. Our augmented data model is mBERT-Aug, which we trained on 6,000 Spanish tweets, and an additional 1,393 back-translated tweets. The additional tweets consist of the English data belonging to the positive class, which were translated back into Spanish using Google Translate API. We also train a bilingual model, by concatenating the output of the two transformer models. We trained this model on both the Spanish and English tweets.

4 Results and Discussion

The bilingual model obtains the best results on the validation data, while mBERT-Aug is the best scoring model on the test data, with a F-1 score of 0.83. Table 1 and Table 2 summarise the results.

We perform experiments after the evaluation period to obtain results on the test data for BERT-base and the bilingual model. We do this to compare the results of all models on the test data. We find that neither the addition of augmented data, nor combining representations in different languages significantly improves the results, with the bilingual model performing better on the validation data, and the mBERT-Aug performing better on the test data. We believe that a reason the BERT-base and bilingual models have lower scores on the test data is due to the quality of the machine translation system that we used whereas the validation data was provided by the task organisers. For example, *Ultima Hora* in Spanish was translated as *last minute*, when it should have been translated as *breaking news* in the context it was used in. Another example is *consellera* which translates to *advisor* was not translated at all in some tweets. While these methods of data augmentation provide a small improvement, fine-tuned pre-trained BERT models are quite robust. Training on parallel corpora gave these models everything that could be extracted through back-translation and bilingual data.

Model	Precision	Recall	F-1
mBERT	0.8407	**0.9347**	0.89
BERT-Base	0.8763	0.8875	0.88
mBert-Aug	0.8826	0.8734	0.88
bilingual	**0.8847**	0.9194	**0.90**

Table 1: ProfNER Task 7a Validation Results

Model	Precision	Recall	F-1
mBERT	0.9538	0.7127	0.82
BERT-Base	0.6620	0.1015	0.18
mBert-Aug	0.9171	**0.7646**	**0.83**
bilingual	**0.9579**	0.6393	0.77

Table 2: ProfNER Task 7a Test Results

5 Conclusion

Our work presents experiments with pre-trained transformer based models to perform binary classification on an imbalanced dataset. We hypothesised that the use of data augmentation and parallel inputs in multiple languages will provide a method of addressing class imbalance (Section 1). However, our experiments showed that neither of these methods are particularly powerful in this regard (Section 4). In the future, we will continue to experiment with other techniques to handle imbalanced classes, such as one-class classification and reinforcement learning-based networks to generate text.

[1] https://github.com/francesita/ProfnerTask7a

References

Jacob Devlin, Ming-Wei Chang, Kenton Lee, and Kristina Toutanova. 2019. BERT: Pre-training of Deep Bidirectional Transformers for Language Understanding.

Alex Fine, Patrick Crutchley, Jenny Blase, Joshua Carroll, and Glen Coppersmith. 2020. Assessing population-level symptoms of anxiety, depression, and suicide risk in real time using NLP applied to social media data. In *Proceedings of the Fourth Workshop on Natural Language Processing and Computational Social Science*, Stroudsburg, PA, USA. Association for Computational Linguistics.

Zhibin Lu, Pan Du, and Jian Yun Nie. 2020. VGCN-BERT: Augmenting BERT with Graph Embedding for Text Classification. In *Lecture Notes in Computer Science (including subseries Lecture Notes in Artificial Intelligence and Lecture Notes in Bioinformatics)*, volume 12035 LNCS, pages 369–382. Springer.

Arjun Magge, Ari Klein, Ivan Flores, Ilseyar Alimova, Mohammed Ali Al-garadi, Antonio Miranda-Escalada, Zulfat Miftahutdinov, Eulàlia Farré-Maduell, Salvador Lima López, Juan M Banda, Karen O'Connor, Abeed Sarker, Elena Tutubalina, Martin Krallinger, Davy Weissenbacher, and Graciela Gonzalez-Hernandez. 2021. Overview of the sixth social media mining for health applications (# smm4h) shared tasks at naacl 2021. In *Proceedings of the Sixth Social Media Mining for Health Applications Workshop & Shared Task*.

Antonio Miranda-Escalada, Eulàlia Farré-Maduell, Salvador Lima López, Vicent Briva-Iglesias, Marvin Agüero-Torales, Luis Gascó-Sánchez, and Martin Krallinger. 2021. The profner shared task on automatic recognition of professions and occupation mentions in social media: systems, evaluation, guidelines, embeddings and corpora. In *Proceedings of the Sixth Social Media Mining for Health Applications Workshop & Shared Task*.

Sam Shleifer. 2019. Low Resource Text Classification with ULMFit and Backtranslation. *CoRR*, abs/1903.09244.

Shoko Wakamiya, Yukiko Kawai, and Eiji Aramaki. 2018. Twitter-Based Influenza Detection After Flu Peak via Tweets With Indirect Information: Text Mining Study. *JMIR Public Health and Surveillance*, 4(3).

Jason Wei and Kai Zou. 2020. EDA: Easy data augmentation techniques for boosting performance on text classification tasks. In *EMNLP-IJCNLP 2019 - 2019 Conference on Empirical Methods in Natural Language Processing and 9th International Joint Conference on Natural Language Processing, Proceedings of the Conference*.

Dylan Whang and Soroush Vosoughi. 2020. Dartmouth CS at WNUT-2020 Task 2: Informative COVID-19 Tweet Classification Using BERT. In *Proceedings of the Sixth Workshop on Noisy User-generated Text (W-NUT 2020)*, pages 480–484, Stroudsburg, PA, USA. Association for Computational Linguistics.

Thomas Wolf, Lysandre Debut, Victor Sanh, Julien Chaumond, Clement Delangue, Anthony Moi, Pierric Cistac, Tim Rault, Remi Louf, Morgan Funtowicz, Joe Davison, Sam Shleifer, Patrick von Platen, Clara Ma, Yacine Jernite, Julien Plu, Canwen Xu, Teven Le Scao, Sylvain Gugger, Mariama Drame, Quentin Lhoest, and Alexander Rush. 2020. Transformers: State-of-the-Art Natural Language Processing. In *Proceedings of the 2020 Conference on Empirical Methods in Natural Language Processing: System Demonstrations*, Stroudsburg, PA, USA. Association for Computational Linguistics.

Yi Zhao, Haixu Xi, and Chengzhi Zhang. 2020. Exploring Occupation Differences in Reactions to COVID-19 Pandemic on Twitter. *Data and Information Management*, 0(0).

IIITN NLP at SMM4H 2021 Tasks: Transformer Models for Classification of Health-Related Tweets

Varad Pimpalkhute, Prajwal Nakhate and **Tausif Diwan**
{pimpalkhutevarad, prajwalnakhate}@gmail.com and tdiwan@iiitn.ac.in
Indian Institute of Information Technology, Nagpur

Abstract

Non-availability of well annotated and balanced datasets is considered as one of the major hurdles in analysing and extracting meaningful information from health-related tweets. Herein, we present transformer based deep learning binary classifiers for distinguishing the health related tweets for the three shared tasks 1a, 4 and 8 of the 6th edition of SMM4H Workshop. We evaluate the different transformer based models viz. RoBERTa (for Task 1a & 4) and BioBERT (for Task 8), along with various dataset balancing techniques. We implement augmentation and sampling techniques so as to improve performance on the imbalanced datasets.

1 Introduction

Twitter has gained a huge popularity among all the social media platforms, especially to share and discuss information related to various aspects of life, including health-related problems. Analysing these health related Tweets and extracting the meaningful information from them is an important task for offering better health related services. With the advancements in sequential deep models, Natural Language Processing (NLP) and underlying processes got benefited from it and effective automation is introduced for the various NLP processes to a great extent. Healthcare research community has developed a keen interest in processing these health related information efficiently using advancements of deep learning. The Sixth Social Media Mining for Health Applications (SMM4H) shared tasks focus on addressing such classic health related problems applied to Twitter micro-corpus (tweets) (Magge et al., 2021).

Our team participated in three different shared binary classification tasks viz. Task 1a, Task 4, and Task 8. Task 1a focuses on distinguishing tweets mentioning adverse drug effects (ADE) from other tweets (NoADE). (O'Connor et al., 2014) focused on the identification of tweets mentioning drugs having potential signals for ADEs. Task 4 focuses on distinguishing tweets mentioning adverse potential outcomes (APO) from other tweets (NoAPO). Task 8 focuses on segregating the tweets containing self-reports (S) of breast cancer from other tweets (NR). The datasets provided for the shared tasks 1a and 8 are highly imbalanced. However, dataset for the shared Task 4 is comparative balanced. Table 1 illustrates the underlying datasets characteristics for the three shred tasks.

Due to the scarcity of users tweeting on health topics, most of the datasets on these topics are highly imbalanced in nature. (Mujtaba et al., 2019) gives a broad overview on the various balancing techniques applied on various medical datasets. (Ebenuwa et al., 2019) demonstrates the effect of strategies such as oversampling and cost-sensitivity on various health-related datasets. (Amin-Nejad et al., 2020; Tayyar Madabushi et al., 2019) presents extension of this work on cost-sensitivity to allow models such as BioBERT and BERT to generalize well on imbalanced datasets. (Liu et al., 2019; Akkaradamrongrat et al., 2019; Padurariu and Breaban, 2019) also present strategies such as text generation techniques, embedded feature extraction methods to generalize the classifier on an imbalanced dataset.

We propose transformer based classification models for the binary classification for all the aforementioned tasks. We especially address the class imbalance in the datasets, for Task 1a and Task 8. We experiment with techniques such as undersampling, oversampling, and data augmentation to address the datasets imbalance for these tasks. The rest of the paper is organized as follows. Section 2 covers the underlying datasets for the three shared tasks, their characteristics, preprocessing details, and sampling techniques to address the inherent imbalance in the dataset. Section 3 presents the classification models for the shared tasks. Re-

Proceedings of the Sixth Social Media Mining for Health Workshop 2021, pages 118–122
June 10, 2021.

Table 1: Dataset characteristics for the shared tasks.

Task	Label	#	Sample Instance
Task1a	ADE	1300	ooh me too! rt @xyle50ul: #schizophrenia #seroquel did not suit me at all. had severe tremors and weight gain..
	NoADE	17000	I need Temazepam and alprazolam.... Is there any doctor can prescribe for me?? :/
Task 4	APO	2922	The LAST thing you wanna do is call my son "slow" or say he's "different than everyone else" because he's a preemie.. Fuck off.
	NoAPO	3565	I don't usually use the term "rainbow baby" myself but I think it's incredibly brave when people share these... https://t.co/jjktHOewDz
Task 8	S	975	@arizonadelight i'm a breast cancer survivor myself so i understand the scare.
	NR	2840	All done, we done for raising awareness, I have a good friend battling this at the moment #breastcancer.

sults and discussions are sketched in the Section 4. Section 5 conclude the paper and presents future research directions.

2 Dataset: Sampling Techniques and Preprocessing

The datasets for the shared tasks were collected in the form of English tweets. The datasets were well annotated for each of the shared tasks. We majorly employ three dataset balancing techniques viz. undersampling, oversampling, and augmentation.

2.1 Sampling Techniques

Under-sampling is performed to balance the data by reducing the instances of the excessive class nearly equal to the rare class. Over-sampling is the approach to duplicate the rare class instances, thus increasing the number of samples of rare class to that of the excess class in the dataset. We achieved this either by addition of tweets of rare class with repetition or using Synthetic Minority Over-Sampling technique (SMOTE) (Bowyer et al., 2011). Performance of these sampling techniques for different ratios of rare to excess class for the dataset of Task 1a on applying RoBERTa model are presented in Figure 1. For our experiments, rare class is ADE / APO / S and excess class is NoADE / NoAPO / NR for three datasets corresponding to three shared tasks.

2.2 Data Augmentation

Data-Augmentation using the nlpaug library (Ma, 2019) is undertaken to balance the datasets. Synthetic data of the rare class is added by generating tweets with different spellings, synonyms, word-embedding, contextual word-embedding of words in-order to have artificial tweets look as natural as real tweets. Data Augmentation is different from Oversampling in the sense that data augmentation adds variations in input text whereas oversampling is not able to change the features of the text.

Figure 1: Sampling performance using RoBERTa model for Task 1a.

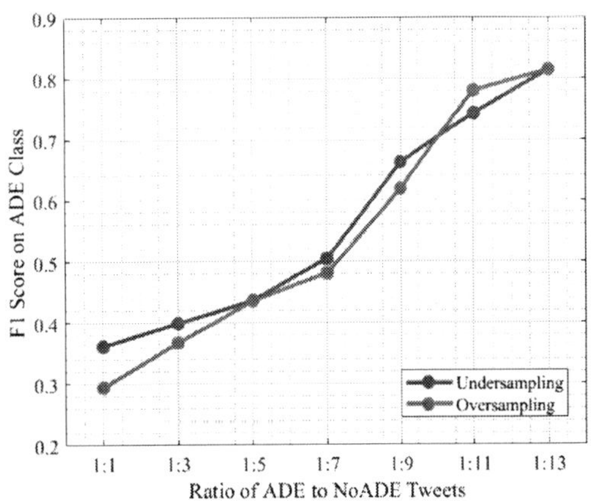

2.3 Pre-processing

Before feeding the dataset to a text classification model, we cleaned and preprocessed the tweets in each of the datasets. For each tweet in the dataset, we normalized usernames and keywords into reserved keywords[1]. We also de-emojized the tweets using the emoji package[2] to replace the emojis with relevant tags. Lastly, we expanded contractions[3]

[1]https://github.com/avian2/unidecode
[2]https://github.com/carpedm20/emoji
[3]https://github.com/kootenpv/contractions

Table 2: Datasets Characteristics for each of the three tasks.

	Task 1a			Task 4			Task 8		
Corpus	**ADE**	**NoADE**	**#**	**NoAPO**	**APO**	**#**	**NR**	**S**	**#**
Train Set	1235	16150	17385	3030	2484	5514	2615	898	3513
Valid Set	65	850	915	535	438	973	225	77	302
Test Set	NA	NA	10000	NA	NA	10000	NA	NA	1204

and lower-cased the text to present the data in a much cleaner format.

3 System Description And Model

We employ transformer based models and their architectural variants for all the shared tasks, along with dataset balancing techniques described in the previous section. For all the tasks, the experiments have been performed using the scikit-learn, Tensorflow[4], PyTorch [5] and Flair (Akbik et al., 2019) frameworks. Table 2 describes the three datasets and their distribution in train, test, and validation sets for training and evaluation of transformer based sequence models.

Figure 2: Proposed model architecture.

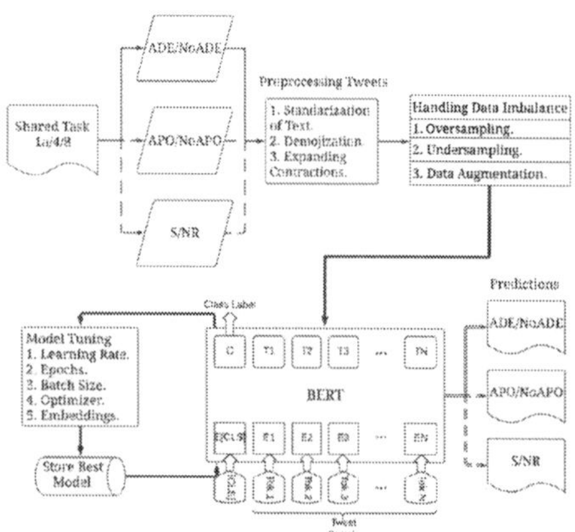

3.1 Classification Model

We mainly experimented with various tranformer languages models such as BERT (Devlin et al., 2018), DistilBert (Sanh et al., 2019), XLNET (Yang et al., 2019), and RoBERTa (Liu et al., 2019). In addition to these routine transformer models, we also experimented on health related architectural variants such as BioBERT (Lee et al., 2019), BERT-Epi (Müller et al., 2020) and BERTweet (Nguyen et al., 2020). Table 3 presents the sample results of all these models for shared task 4. In the subsequent section, we demonstrated the results for the best preforming transformer models for each of the shared tasks. Furthermore, we penalized the loss of the rare class with a loss weight two times the original loss weight. We kept the loss weight for the excess class as it is. We experimented each of the models on four different versions of the underlying dataset: Original, Undersampled, Oversampled and Augmented. The architecture of our proposed system is illustrated in Figure 2.

Table 3: Comparative results of various transformer based models for the shared task 4.

Architecture	xLR ($\times 10^{-6}$)	F1	Prec	Recall
BERT	10	0.872	0.843	0.902
BERTweet	10	0.899	0.896	0.906
DistilBERT	50	0.835	0.839	0.831
RoBERTa	6	**0.924**	0.897	**0.952**
XLNET	5	0.903	**0.922**	0.886
BioBERT	5	0.874	0.859	0.890

3.2 Hyperparamter Tuning

All the experiments have been performed on Flair Framework. We tried various ensemble of models – where, there were three models in each ensemble – but, this didn't draw good results on the validation set, thus, we choose the final model as a single transformer language model. Ensembling didn't work well as majority of the incorrectly predicted samples were predicted incorrectly by most of the models in the ensemble. For Task 1 and Task 4, we choose the final transformer model as RoBERTa, and for Task 8 we made use of a health related model trained on COVID19 related tweets – BioBERT. We experimented with various hyperparamter settings such as learning rate, learning rate decay, early stopping, varying batch size, and number of epochs. Based on the various experiments, we settled that the learning rate in the range of 0.000006 - 0.00001, batch size of 8, patience of 2

[4]https://www.tensorflow.org/
[5]https://pytorch.org/

and 3 epochs of training gave the best performance on the models. The performance was measured across standard metrics such as precision and recall, with the final determining metric being the harmonic mean of precision and recall (F1-score) for the rare classes.

4 Results & Discussions

All the experiments were performed on an Intel core i5 CPU @2.50GHz, 8GB RAM machine having 4 logical cores. The task wise results can be presented as follows:

4.1 Task 1a: Adverse Drug Effect Mentions.

Table 4: Task 1a using RoBERTa (Learning Rate = 1×10^{-5}, Epochs = 3).

Validation set			
Dataset	**F1**	**Precision**	**Recall**
Undersampled	0.5048	0.5561	0.4623
Oversampled	0.4361	0.4186	0.4553
Original	0.8136	**0.9057**	0.7385
Augmented	**0.8433**	0.8209	**0.8572**
Test set			
Dataset	**F1**	**Precision**	**Recall**
Original	0.3	0.473	0.217
Augmented	0.4	0.405	0.401
Median	**0.44**	**0.505**	**0.409**

As we know, Task 1 is a highly imbalanced dataset with the ratio of ADE to NoADE tweets being about 1:13. Table 4 presents the metrics on the validation as well as test data for Task 1a. As it can be observed, RoBERTa shows the best performance on Augmented Dataset. Undersampling results in underfitting the training model whereas oversampling results in model overfitting. The probable reason behind this is the sparse ADE samples present in the dataset for the shared Task 1a. In contrast, data augmentation results in increasing variations in the training dataset, thus, we are able to generalize well as compared to the original dataset.

4.2 Task 4: Self-reporting Adverse Pregnancy Outcome.

Similar to Section 4.1, RoBERTa model shows the best performance on the validation set for the shared Task 4 also, represented using Table 5. As Task 4 dataset was comparatively balanced, there was little motivation for using sampling techniques on the dataset. Surprisingly, augmenting the data couldn't draw better F1 score.

4.3 Task 8: Breast Cancer Self-reports.

Task 8 is also an imbalanced dataset with the ratio of Self-Reports to Non-Relevant Tweets being about 1:3. Thus, similar to Section 4.1, we experiment with all the four variations of the dataset. The metrics on the validation and test data are presented in Table 6. It can be seen that the model with the best performance is on the augmented dataset. As the imbalance in Task 8 was significantly lower than that in Task 1, we observe better results for this task.

Table 5: Task 4 using RoBERTa (Learning Rate = 6×10^{-6}, Epochs = 5).

Validation set			
Dataset	**F1**	**Precision**	**Recall**
Original	0.9437	0.9251	0.9631
Augmented	0.9279	0.9028	0.9543
Test set			
Dataset	**F1**	**Precision**	**Recall**
Original	**0.93**	0.9149	**0.9412**
Augmented	0.92	0.8919	0.948
Median	0.925	**0.9183**	0.9234

Table 6: Task 8 using BioBERT (Learning Rate = 5×10^{-6}, Epochs = 10).

Validation set			
Dataset	**F1**	**Precision**	**Recall**
Undersampled	0.8182	0.7273	0.9351
Oversampled	0.828	0.8125	0.8442
Original	0.8707	**0.9143**	0.8313
Augmented	**0.8947**	0.9067	**0.8831**
Test set			
Dataset	**F1**	**Precision**	**Recall**
Original	0.83	0.8441	0.8216
Augmented	0.84	**0.8706**	0.8084
Median	**0.85**	0.8701	**0.8377**

5 Conclusions

We proposed a text classification pipeline while also making an attempt to handle dataset imbalance corresponding to three different shared tasks in SMM4H'21 (Magge et al., 2021). We conclude that data augmentation gives best performance on highly imbalanced datasets. Moreover, augmentation provides better results in case of comparatively balanced datasets. As part of future work, additional experiments are planned to further analyze strategies to improve the performance of the model on the dataset.

References

Alan Akbik, Tanja Bergmann, Duncan Blythe, Kashif Rasul, Stefan Schweter, and Roland Vollgraf. 2019. FLAIR: An easy-to-use framework for state-of-the-art NLP. In *Proceedings of the 2019 Conference of the North American Chapter of the Association for Computational Linguistics (Demonstrations)*, pages 54–59, Minneapolis, Minnesota. Association for Computational Linguistics.

S. Akkaradamrongrat, P. Kachamas, and S. Sinthupinyo. 2019. Text generation for imbalanced text classification. In *2019 16th International Joint Conference on Computer Science and Software Engineering (JCSSE)*, pages 181–186.

Ali Amin-Nejad, Julia Ive, and Sumithra Velupillai. 2020. Exploring transformer text generation for medical dataset augmentation. In *Proceedings of the 12th Language Resources and Evaluation Conference*, pages 4699–4708, Marseille, France. European Language Resources Association.

Kevin W. Bowyer, Nitesh V. Chawla, Lawrence O. Hall, and W. Philip Kegelmeyer. 2011. SMOTE: synthetic minority over-sampling technique. *CoRR*, abs/1106.1813.

Jacob Devlin, Ming-Wei Chang, Kenton Lee, and Kristina Toutanova. 2018. BERT: pre-training of deep bidirectional transformers for language understanding. *CoRR*, abs/1810.04805.

S. H. Ebenuwa, M. S. Sharif, M. Alazab, and A. Al-Nemrat. 2019. Variance ranking attributes selection techniques for binary classification problem in imbalance data. *IEEE Access*, 7:24649–24666.

Jinhyuk Lee, Wonjin Yoon, Sungdong Kim, Donghyeon Kim, Sunkyu Kim, Chan Ho So, and Jaewoo Kang. 2019. Biobert: a pre-trained biomedical language representation model for biomedical text mining. *CoRR*, abs/1901.08746.

H. Liu, M. Zhou, and Q. Liu. 2019. An embedded feature selection method for imbalanced data classification. *IEEE/CAA Journal of Automatica Sinica*, 6(3):703–715.

Yinhan Liu, Myle Ott, Naman Goyal, Jingfei Du, Mandar Joshi, Danqi Chen, Omer Levy, Mike Lewis, Luke Zettlemoyer, and Veselin Stoyanov. 2019. Roberta: A robustly optimized BERT pretraining approach. *CoRR*, abs/1907.11692.

Edward Ma. 2019. Nlp augmentation. https://github.com/makcedward/nlpaug.

Arjun Magge, Ari Klein, Ivan Flores, Ilseyar Alimova, Mohammed Ali Al-garadi, Antonio Miranda-Escalada, Zulfat Miftahutdinov, Eulàlia Farré-Maduell, Salvador Lima López, Juan M Banda, Karen O'Connor, Abeed Sarker, Elena Tutubalina, Martin Krallinger, Davy Weissenbacher, and Graciela Gonzalez-Hernandez. 2021. Overview of the sixth social media mining for health applications (# smm4h) shared tasks at naacl 2021. In *Proceedings of the Sixth Social Media Mining for Health Applications Workshop & Shared Task*.

Ghulam Mujtaba, Liyana Shuib, Norisma Idris, Wai Lam Hoo, Ram Gopal Raj, Kamran Khowaja, Khairunisa Shaikh, and Henry Friday Nweke. 2019. Clinical text classification research trends: Systematic literature review and open issues. *Expert Systems with Applications*, 116:494–520.

Martin Müller, Marcel Salathé, and Per E Kummervold. 2020. Covid-twitter-bert: A natural language processing model to analyse covid-19 content on twitter. *arXiv preprint arXiv:2005.07503*.

Dat Quoc Nguyen, Thanh Vu, and Anh Tuan Nguyen. 2020. BERTweet: A pre-trained language model for English Tweets. In *Proceedings of the 2020 Conference on Empirical Methods in Natural Language Processing: System Demonstrations*.

Karen O'Connor, Pranoti Pimpalkhute, Azadeh Nikfarjam, Rachel Ginn, , Karen Smith, and Graciela Gonzalez. 2014. Pharmacovigilance on twitter? mining tweets for adverse drug reactions. pages 924–33.

Cristian Padurariu and Mihaela Elena Breaban. 2019. Dealing with data imbalance in text classification. *Procedia Computer Science*, 159:736–745. Knowledge-Based and Intelligent Information & Engineering Systems: Proceedings of the 23rd International Conference KES2019.

Victor Sanh, Lysandre Debut, Julien Chaumond, and Thomas Wolf. 2019. Distilbert, a distilled version of BERT: smaller, faster, cheaper and lighter. *CoRR*, abs/1910.01108.

Harish Tayyar Madabushi, Elena Kochkina, and Michael Castelle. 2019. Cost-sensitive BERT for generalisable sentence classification on imbalanced data. In *Proceedings of the Second Workshop on Natural Language Processing for Internet Freedom: Censorship, Disinformation, and Propaganda*, pages 125–134, Hong Kong, China. Association for Computational Linguistics.

Zhilin Yang, Zihang Dai, Yiming Yang, Jaime G. Carbonell, Ruslan Salakhutdinov, and Quoc V. Le. 2019. Xlnet: Generalized autoregressive pretraining for language understanding. *CoRR*, abs/1906.08237.

OCHADAI at SMM4H-2021 Task 5: Classifying self-reporting tweets on potential cases of COVID-19 by ensembling pre-trained language models

Ying Luo[1], Lis Kanashiro Pereira[1], and Ichiro Kobayashi[1]

[1]Ochanomizu University, Japan

1 Introduction

Since the outbreak of coronavirus at the end of 2019, there have been numerous studies on coronavirus in the NLP arena. Meanwhile, Twitter has been a valuable source of news and a public medium for the conveyance of information and personal expression. This paper describes the system developed by the Ochadai team for the Social Media Mining for Health Applications (SMM4H) 2021 Task 5, which aims to automatically distinguish English tweets that self-report potential cases of COVID-19 from those that do not. We proposed a model ensemble that leverages pre-trained representations from COVID-Twitter-BERT (Müller et al., 2020), RoBERTa (Liu et al., 2019), and Twitter-RoBERTa (Glazkova et al., 2021). Our model obtained F1-scores of 76% on the test set in the evaluation phase, and 77.5% in the post-evaluation phase.

2 System Overview

In this section, we overview the pre-processing steps, pre-trained language models and training prodedure used by our system.

2.1 Text pre-processing

We follow (Müller et al., 2020) for pre-processing the dataset. First, we lowercase the text. Then, we replace user tags (e.g. @ScottGottliebMD) with the token "@USER", and replace urls with the token "URL". All unicode emoticons are replaced with textual ASCII representations (e.g. dog for ) using the Python emoji library [1]. We also remove the unicode symbols (e.g. & for &), control characters and accented characters (e.g. shyapu for shyápu).

[1]https://pypi.org/project/emoji/

2.2 Pre-trained Models

We mainly experimented with three transformer-based pre-trained language models as follows:

COVID-Twitter-BERT (CT-BERT) (Müller et al., 2020): This is a $\text{BERT}_{\text{LARGE}}$ model trained on a large corpus of Twitter messages on the topic of COVID-19, collected during the period from January 12 to April 16, 2020.

$\textbf{RoBERTa}_{\textbf{LARGE}}$ (Liu et al., 2019): We use the $\text{RoBERTa}_{\text{LARGE}}$ models released by the authors. Similar to $\text{BERT}_{\text{LARGE}}$, $\text{RoBERTa}_{\text{LARGE}}$ consists of 24 transformer layers, 16 self-attention heads per layer, and a hidden size of 1024.

Twitter-RoBERTa (Glazkova et al., 2021): This is a $\text{RoBERTa}_{\text{BASE}}$ model pre-trained on a large corpus of English tweets. This corpus includes tweets from 2020, possibly covering the COVID-19 topic as well.

2.3 Training Procedure

We fine-tuned each pre-trained language model on the training set with 5-fold cross-validation. We ran each model using three different random seeds, and selected the best performing model on the validation set or averaged the prediction probabilities obtained after softmax. Then, we further combined the outputs of the models generated by each fold by again taking an average of the prediction probabilities obtained after softmax. We also experimented on max-voting on the predicted labels.

We further experimented on ensembling CT-BERT, RoBERTa-large, and Twitter-RoBERTa.

Proceedings of the Sixth Social Media Mining for Health Workshop 2021, pages 123–125
June 10, 2021.

Index	Training Models	Ensemble			Cross Validation	Ensemble Method		F1-Score	
						average probability	max voting	Validation	Test
1	Covid-Twitter-BERT							78.00	
2	Covid-Twitter-BERT				✔		✔	78.40	
3	Covid-Twitter-BERT	★	★		✔			92.00	
4	Covid-Twitter-BERT	★	★	★	✔	✔		79.00	
5	Twitter-RoBERTa-base							86.00	
6	Twitter-RoBERTa-base				✔		✔	92.00	
7	Twitter-RoBERTa-base				✔	✔		**93.00**	
9	RoBERTa-large							83.00	
10	RoBERTa-large				✔		✔	92.00	
11	RoBERTa-large				✔	✔		**93.00**	
12	*Twitter-RoBERTa-base+Covid-Twitter-BERT+RoBERTa-large*	★	★			✔		94.00	
13	*Twitter-RoBERTa-base+Covid-Twitter-BERT+RoBERTa-large*	★		★	✔	✔		77.00	
14	Ensemble 1*	○					✔	**97.00**	76.00
15	Ensemble 2 ‡		○			✔		93.00	**77.50**
16	Ensemble 3 ‡			○			✔	93.00	76.67

Table 1: Comparison of different text encoders and different ensemble methods. Best results are highlighted in bold. ★ indicates each model that was used in the Ensemble 1, Ensemble 2, and Ensemble 3 models, respectively indicated in each column by ○. * indicates the models submitted during the evaluation phase, and ‡ indicates the models submitted during the post-evaluation phase.

3 Experiments

3.1 Implementation Details

In this work, we used the PyTorch implementation released by huggingface[2] of RoBERTa$_{\text{LARGE}}$, Covid-Twitter-BERT, and Twitter-RoBERTa. We used AdamW as our optimizer, with a learning rate in the range $\in \{9 \times 10^{-6}, 1 \times 10^{-5}, 2 \times 10^{-5}\}$ and a batch size $\in \{16, 32\}$. The maximum number of epochs was set to $\in \{5, 10\}$. A linear learning rate decay schedule with warm-up over 0.01 was used. All the texts were tokenized using wordpieces and were chopped to spans no longer than 512 tokens.

The performance of the models were measured in terms of F1-score, and the model with the highest performance on the validation set was selected.

3.2 Main Results and Analysis

Our results are shown in Table 1. First, we observe that performing cross-validation and averaging the results of each fold yields to better performance on the validation set than max-voting. For instance, the Covid-Twitter-BERT could improve the F1-score from 78.00% to 78.40% to 79.00% on the validation set in lines 1,2,4 of the table. The same tendency can be observed on the F1-score of the validation set in the Twitter-RoBERTa-base (from 86% to 92% and 93% in lines 5,6,7 of the table) and RoBERTa-large (from 83% to 92% and 93% in lines 9,10,11 of the table) models. Moreover, another observation is that combining the outputs of models by taking an average of the prediction probabilities obtained after softmax instead of max-voting on the predicted labels leads to higher performance on the validation set. For instance, the improved F1-score on validation set was observed from the table in the Twitter-RoBERTa-base (from 92% to 93% in lines 6,7 of the table) and RoBERTa-large (from 92% to 93% in lines 10,11 of the table) models.

Finally, ensembling different pre-trained models leads to better performance on the test set. For instance, the Covid-Twitter-BERT model submitted in the evaluation phase obtained an F1-score of 69% which is not referred in the table, while the Ensemble 1, Ensemble 2, and Ensemble 3 models obtained F1-scores of 76%, 77.5%, and 76.66%, respectively.

4 Conclusion

We presented the Ochadai system submitted to the SMM4H-2021 Task 5. We proposed an ensemble model that leverages pre-trained representations from COVID-Twitter-BERT (Müller et al., 2020), RoBERTa (Liu et al., 2019), and Twitter-RoBERTa (Glazkova et al., 2021). Our best performing model obtained an F1-score of 77.5%. In future efforts, we plan to further improve our model by exploring other pre-trained language models and ensemble techniques.

[2] https://huggingface.co/models

References

Anna Glazkova, Maksim Glazkov, and Timofey Trifonov. 2021. g2tmn at constraint@aaai2021: Exploiting ct-bert and ensembling learning for covid-19 fake news detection.

Yinhan Liu, Myle Ott, Naman Goyal, Jingfei Du, Mandar Joshi, Danqi Chen, Omer Levy, Mike Lewis, Luke Zettlemoyer, and Veselin Stoyanov. 2019. Roberta: A robustly optimized bert pre-training approach.

Martin Müller, Marcel Salathé, and Per E Kummervold. 2020. Covid-twitter-bert: A natural language processing model to analyse covid-19 content on twitter.

PAII-NLP at SMM4H 2021: Joint Extraction and Normalization of Adverse Drug Effect Mentions in Tweets

Zongcheng Ji, Tian Xia and **Mei Han**
PAII Inc.
3000 El Camino Real, Palo Alto, CA 94306
{jizongcheng, SummerRainET2008, hanmei613}@gmail.com

Abstract

This paper describes our system developed for the subtask 1c of the sixth Social Media Mining for Health Applications (SMM4H) shared task in 2021. The aim of the subtask is to recognize the adverse drug effect (ADE) mentions from tweets and normalize the identified mentions to their mapping MedDRA preferred term IDs. Our system is based on a neural transition-based joint model, which is to perform the recognition and normalization simultaneously. Our final two submissions outperform the average F1 by 1-2%.

1 Introduction

With the popularity of social media such as Twitter, people often publish messages online in regard to their health such as the information related to the adverse drug effects (ADEs). Mining such type of information from social media is helpful for pharmacological post-marketing surveillance and monitoring. The aim of the sixth Social Media Mining for Health Applications (SMM4H) shared task in 2021 (Magge et al., 2021) is to mining such invaluable health information from social media. We participate in the subtask 1c of SMM4H 2021, which is to recognize the ADE mentions from tweets and normalize the identified mentions to their mapping MedDRA [1] preferred term IDs.

2 Task and Data Description

We give the formal definition of the end-to-end task. Briefly, given a tweet x published by a user, and a knowledge base (KB, i.e., MedDRA) which consists of a set of concepts, the goal of the task is to identify all the ADE mentions $M = \{m_1, m_2, ..., m_{|M|}\}$ in x and to link each of the identified mention m_i to the mapping MedDRA preferred term ID e_i in KB, $m_i \rightarrow e_i$. If there is no mapping concept in KB for m_i, then $m_i \rightarrow NIL$, where NIL denotes that m_i is unlinkable.

Table 1: Overall statistics of the dataset.

	#tweets	#mentions	#unique concepts
trn	17,375	1,706	317
dev	915	86	57
tst	10,984	-	-

Table 1 shows the statistics of the dataset provided by the organizers. We use the training (trn) and development (dev) sets to build our system and submit the predictions on the testing (tst) set.

We use MedDRA v21.1 as the KB, which consists 25,463 unique preferred term IDs.

3 The Approach

Preprocessing. We preprocess all the tweets with the following steps (Ji et al., 2016): 1) tokenize the tweets with whitespace and punctuations; 2) lowercase the tokens; 3) replace the urls with "httpurl"; 4) replace the @user with "username"; 5) replace the escape characters with their original form (e.g., & $\rightarrow$ &).

We preprocess all the mentions and concepts in KB with the following steps (Ji et al., 2020): 1) replace the numerical words to their corresponding Arabic numerals (e.g., one / first / i / single $\rightarrow$ 1); 2) tokenize the mentions and concepts with whitespace and punctuations; 3) remove the punctuations; 4) lowercase the tokens.

Neural Transition-based Joint Model. We cast the end-to-end task as a sequence labeling task and convert the whole task as an action sequence prediction task. We follow previous studies of applying Neural Transition-based Model for named entity recognition (NER) (Lample et al., 2016; Wang et al., 2018) with SHIFT, OUT, REDUCE, SEGMENT actions for the recognition purpose and further extend the model by adding LINKING actions for the normalization purpose.

[1] https://www.meddra.org/

Proceedings of the Sixth Social Media Mining for Health Workshop 2021, pages 126–127
June 10, 2021.

Table 2: Results on the development set.

	Precision	Recall	F1
Submission 1	0.623	0.545	0.582
Submission 2	0.570	0.557	0.563

Table 3: Results on the test set.

	Precision	Recall	F1
Submission 1	0.331	0.179	0.230
Submission 2	0.317	0.196	0.240
Average	0.231	0.218	0.220

Input Representation We represent each token x_i in a tweet x by concatenating its character-level word representation, non-contextual word representation, and contextual word representation:

$$x_i = [v_i^{char}; v_i^{w}; ELMo_i] \quad (1)$$

where v_i^{char} denotes its character-level word representation learned by using a CNN network (Ma and Hovy, 2016), v_i^{w} denotes its non-contextual word representation initialized with Glove (Pennington et al., 2014) embeddings, which is pre-trained on a large-scale Twitter corpus of two billion tweets, and $ELMo_i$ denotes its contextual word representation initialized with ELMo (Peters et al., 2018).

Search and Training For efficient decoding, a widely-used greedy search algorithm (Lample et al., 2016; Wang et al., 2018) is adopted to minimize the negative log-likelihood of the local action classifier, *i.e.,* to minimize the cross-entropy loss between the output distribution with the gold-standard distribution:

$$\mathcal{L}(\theta) = -\sum_t log\ p(a_t|r_t) \quad (2)$$

where θ denotes all the parameters in this model.

4 Results and Conclusions

We submit the following two results with two different strategies:

- **Submission 1**: single model result with the neural transition-based joint model.
- **Submission 2**: voting result with 5 best single model results.

We report the Precision, Recall and F1 for each ADE extracted where the spans overlap either entirely or partially AND each span is normalized to the correct MedDRA preferred term ID.

Table 2 and 3 show the evaluation results on the development and test sets, respectively. Average denotes the arithmetic median of all submissions made by all the teams participate the end-to-end subtask. Results show that the proposed method outperform the average F1 by 1-2%.

In the future, we will further tune the model and explore other popular contextual word representations learned from BERT (Devlin et al., 2018).

References

Jacob Devlin, Ming-Wei Chang, Kenton Lee, and Kristina Toutanova. 2018. BERT: Pre-training of deep bidirectional transformers for language understanding. *arXiv preprint arXiv:1810.04805.*

Zongcheng Ji, Aixin Sun, Gao Cong, and Jialong Han. 2016. Joint Recognition and Linking of Fine-Grained Locations from Tweets. In *WWW*, pages 1271–1281.

Zongcheng Ji, Qiang Wei, and Hua Xu. 2020. BERT-based Ranking for Biomedical Entity Normalization. In *AMIA 2020 Informatics Summit*, pages 269–277.

Guillaume Lample, Miguel Ballesteros, Sandeep Subramanian, Kazuya Kawakami, and Chris Dyer. 2016. Neural Architectures for Named Entity Recognition. In *NAACL*, pages 260–270, San Diego, California. Association for Computational Linguistics.

Xuezhe Ma and Eduard Hovy. 2016. End-to-end Sequence Labeling via Bi-directional LSTM-CNNs-CRF. In *ACL*, pages 1064–1074, Berlin, Germany. Association for Computational Linguistics.

Arjun Magge, Ari Klein, Ivan Flores, Ilseyar Alimova, Mohammed Ali Al-garadi, Antonio Miranda-Escalada, Zulfat Miftahutdinov, Eulàlia Farré-Maduell, Salvador Lima López, Juan M Banda, Karen O'Connor, Abeed Sarker, Elena Tutubalina, Martin Krallinger, Davy Weissenbacher, and Graciela Gonzalez-Hernandez. 2021. Overview of the Sixth Social Media Mining for Health Applications (# SMM4H) Shared Tasks at NAACL 2021. In *Proceedings of the Sixth Social Media Mining for Health Applications Workshop & Shared Task.*

Jeffrey Pennington, Richard Socher, and Christopher Manning. 2014. Glove: Global vectors for word representation. In *EMNLP*, pages 1532–1543.

Matthew E Peters, Mark Neumann, Mohit Iyyer, Matt Gardner, Christopher Clark, Kenton Lee, and Luke Zettlemoyer. 2018. Deep Contextualized Word Representations. In *NAACL-HLT*, pages 2227–2237.

Bailin Wang, Wei Lu, Yu Wang, and Hongxia Jin. 2018. A Neural Transition-based Model for Nested Mention Recognition. In *EMNLP*, pages 1011–1017. Association for Computational Linguistics.

Assessing multiple word embeddings for named entity recognition of professions and occupations in health-related social media

Vasile Păiș and **Maria Mitrofan**
Research Institute for Artificial Intelligence "Mihai Drăgănescu", Romanian Academy
Casa Academiei, Calea 13 Septembrie nr. 13, sector 5, Bucureşti, ROMÂNIA
vasile,maria@racai.ro

Abstract

This paper presents our contribution to the ProfNER shared task. Our work focused on evaluating different pre-trained word embedding representations suitable for the task. We further explored combinations of embeddings in order to improve the overall results.

1 Introduction

The ProfNER task (Miranda-Escalada et al., 2021b), part of the SMM4H workshop and shared task (Magge et al., 2021) organized at NAACL 2021, focused on identification of professions and occupations from health-relevant Twitter messages written in Spanish. It offered two sub-tasks: a) a binary classification task, deciding if a particular tweet contains a mention of an occupation, given the context, and b) extracting the actual named entities, by specifying the entity type, start and end offset as well as the actual text span.

Habibi et al. (2017) have shown that domain specific embeddings have an impact on the performance of a NER system. The ProfNER task is at a confluence between multiple domains. The classification sub-task suggests that tweets will actually contain not only health-related messages but probably also more general domain messages. However, the second task focuses on the analysis of health-related messages. Finally, social media can be regarded as a domain in itself. Therefore, our system was constructed on the assumption that word embeddings from multiple domains (general, health-related, social media) will have different impact on the performance of a NER system. We evaluated different pre-trained embeddings alone and in combination, as detailed in the next section.

Our interest for the task stemmed from our involvement with the CURLICAT[1] project for the CEF AT action, where NER in different domains (including health-related) is needed. Additionally, pre-trained word embeddings for Romanian language, such as Paiș and Tufiș (2018), are considered for suitability in different tasks within the European Language Equality (ELE)[2] project.

2 System description and results

We used a recurrent neural network model based on LSTM cells with token representation using pre-trained word embeddings and additional character embeddings, computed on the fly. The actual prediction is performed by a final CRF layer. For the implementation we used the NeuroNER[3] (Dernoncourt et al., 2017) package.

We considered the two sub-tasks to be intertwined. If a correct classification is given for the first sub-task, then this can be used in the second task to guide the NER process to execute only on the classified documents. However, also the reverse can be applied. A document containing correctly identified entities for the second sub-task should be classified as belonging to the domain of interest. We employed the second approach and first performed NER and then used this information for classification.

For the purposes of the NER sub-task we considered the following word embedding representations: Spanish Medical Embeddings[4] (Soares et al., 2019), Wikipedia Embeddings[5] (Mikolov et al., 2018), Twitter Embeddings[6] (Miranda-Escalada et al., 2021a). These were generated using the FastText toolkit (Bojanowski et al., 2017) and contain floating point vectors of dimension 300. The Spanish Medical Embeddings offers three variants

[1] https://curlicat-project.eu/

[2] http://www.european-language-equality.eu

[3] http://neuroner.com/

[4] https://zenodo.org/record/3744326#.YEbu950zZPZ

[5] https://fasttext.cc/docs/en/english-vectors.html

[6] https://zenodo.org/record/4449930#.YEbwUp0zZPY

Proceedings of the Sixth Social Media Mining for Health Workshop 2021, pages 128–130
June 10, 2021.

Representation	P	R	F1
Medical	83.70	69.43	75.90
Twitter	82.92	71.58	76.83
Wiki	80.63	74.19	77.28
Twitter+Wiki	79.93	72.20	75.87
Twitter+Wiki (all)	81.90	72.96	77.17
Wiki+Twitter	80.86	75.27	77.96
Wiki+Twitter+Med	**83.84**	**75.73**	**79.58**

Table 1: Results of different word embeddings and combinations on the validation set for the NER subtask

Representation	P	R	F1
Medical	**92.38**	86.37	89.27
Twitter	92.05	87.42	89.68
Wiki	90.08	**89.52**	89.80
Twitter+Wiki	90.83	89.31	**90.06**
Twitter+Wiki (all)	91.67	87.63	89.60
Wiki+Twitter	89.68	89.31	89.50
Wiki+Twitter+Med	91.18	88.89	90.02

Table 2: Results of different word embeddings and combinations on the validation set for the Classification subtask

Representation	NER F1	Classification F1
Medical	73.60	86.43
Twitter	74.60	88.04
Wiki	75.40	88.72
Twitter+Wiki	76.20	**88.98**
Wiki+Twitter	75.70	88.38
Twitter+Wiki (all)	75.30	88.24
Wiki+Twitter+Med	**78.50**	88.81

Table 3: Results of different word embeddings and combinations on the test set for both subtasks

based on the SciELO[7] database of scientific articles, filtered Wikipedia (comprising the categories Pharmacology, Pharmacy, Medicine and Biology) and a reunion of the two datasets. For all three corpora, representations are available using CBOW and Skip-Gram algorithms, as described in Bojanowski et al. (2017). However we only used the Skip-Gram variants for our experiments, due to the availability of this type of pre-trained vectors for all the considered representations.

We first experimented with individual representations and then began experimenting with sets of two embeddings concatenated. For the words present in the first considered embedding we added the corresponding vector from the second embedding or a zero vector. This provided an input vector of size 600 (resulting from concatenating two vectors of size 300 each), which required the adaptation of the network size accordingly. Additionally we considered a full combination of Twitter and Wikipedia embeddings, placing zero-valued vectors if words were also missing from the first embedding. A final experiment was conducted on a concatenation of 3 embeddings (total vector size 900). Results on the validation set are presented in Table 1 and Table 2, while results on the test set are in Table 3.

[7]https://scielo.org/

Given the word embeddings size (300, 600 and 900, depending on the experiment), the neural network was changed to have a token LSTM hidden layer of the same size. Other hyper-parameters, common to all experiments, are: character embedding of size 25, learning rate of 0.005, dropout rate 0.5 and early stopping if no improvement was achieved for 10 epochs.

Experiments show that given the recurrent neural architecture used, the best single embeddings results, considering overall F1 score, for both subtasks are provided by the Wikipedia embeddings (a general domain representation). However, the Medical Embeddings seem to achieve higher precision. Considering the NER task, the combination of Wikipedia and Twitter achieves the highest F1 from the two embeddings experiments, while the three embeddings combination provides the final best score.

For the first subtask we used the predictions given by a NER model and considered a tweet with at least one recognized entity to belong to the domain required by the subtask. In order to improve recall we further extracted a list of professions from the training set of the NER subtask. This list was filtered and we removed strings that tend to appear many times in tweets labelled "0" in the training set belonging to the classification task. The filtered list was applied in addition to the NER information and texts that had no extracted entities were labelled "1" if they contained any string from the list. This allowed us to further increase the classifier's performance.

3 Conclusions

We investigated the suitability of different representations for analysing text from the health domain in social media, particularly Twitter messages.

Contrary to our initial assumption, a general domain representation (Wikipedia based) provided the best NER results, considering single representations. However, a combination of word embeddings achieved the highest F1 score. For both validation and test datasets, the best models considering F1 are a combination of Twitter and Wikipedia for the NER task and a combination of all three models for the classification task. We consider this to be explainable by the characteristic of social media messages where people do not necessarily restrict their language to in-domain vocabulary (in this case health related) but rather mix in-domain messages with out-of-domain messages or even combine in the same message sentences from multiple domains.

Acknowledgements

This research was supported by the EC grant INEA/CEF/ICT/A2018/28592472 for the Action No: 2019-EU-IA-0034 entitled "Curated Multilingual Language Resources for CEF.AT" (CURLICAT) and by the European Language Equality (ELE) project.

References

Piotr Bojanowski, Edouard Grave, Armand Joulin, and Tomas Mikolov. 2017. Enriching word vectors with subword information. arXiv:1607.04606.

Franck Dernoncourt, Ji Young Lee, and Peter Szolovits. 2017. NeuroNER: an easy-to-use program for named-entity recognition based on neural networks. In *Proceedings of the 2017 Conference on Empirical Methods in Natural Language Processing: System Demonstrations*, pages 97–102, Copenhagen, Denmark. Association for Computational Linguistics.

Maryam Habibi, Leon Weber, Mariana Neves, David Luis Wiegandt, and Ulf Leser. 2017. Deep learning with word embeddings improves biomedical named entity recognition. *Bioinformatics*, 33(14):i37–i48.

Arjun Magge, Ari Klein, Ivan Flores, Ilseyar Alimova, Mohammed Ali Al-garadi, Antonio Miranda-Escalada, Zulfat Miftahutdinov, Eulàlia Farré-Maduell, Salvador Lima López, Juan M Banda, Karen O'Connor, Abeed Sarker, Elena Tutubalina, Martin Krallinger, Davy Weissenbacher, and Graciela Gonzalez-Hernandez. 2021. Overview of the sixth social media mining for health applications (# smm4h) shared tasks at naacl 2021. In *Proceedings of the Sixth Social Media Mining for Health Applications Workshop & Shared Task.*

Tomas Mikolov, Edouard Grave, Piotr Bojanowski, Christian Puhrsch, and Armand Joulin. 2018. Advances in pre-training distributed word representations. In *Proceedings of the International Conference on Language Resources and Evaluation (LREC 2018).*

Antonio Miranda-Escalada, Marvin Aguero, and Martin Krallinger. 2021a. Spanish covid-19 twitter embeddings in fasttext. *Zenodo.*

Antonio Miranda-Escalada, Eulàlia Farré-Maduell, Salvador Lima López, Luis Gascó-Sánchez, Vicent Briva-Iglesias, Marvin Agüero-Torales, and Martin Krallinger. 2021b. The profner shared task on automatic recognition of occupation mentions in social media: systems, evaluation, guidelines, embeddings and corpora. In *Proceedings of the Sixth Social Media Mining for Health Applications Workshop & Shared Task.*

Vasile Pais and Dan Tufiș. 2018. Computing distributed representations of words using the corola corpus. *Proceedings of the Romanian Academy, Series A*, 19(2):403–410.

Felipe Soares, Marta Villegas, Aitor Gonzalez-Agirre, Martin Krallinger, and Jordi Armengol-Estapé. 2019. Medical word embeddings for Spanish: Development and evaluation. In *Proceedings of the 2nd Clinical Natural Language Processing Workshop*, pages 124–133, Minneapolis, Minnesota, USA. Association for Computational Linguistics.

Fine-Tuning Transformers for Identifying Self-Reporting Potential Cases and Symptoms of COVID-19 in Tweets

Max Fleming[1] **Priyanka Dondeti**[2] **Caitlin N. Dreisbach**[3] **Adam Poliak**[2,3]
Johns Hopkins University[1] Barnard College[2]
Data Science Institute, Columbia University[3]
mflemi21@jhu.edu, {pdd2112,apoliak}@barnard.edu, c.dreisbach@columbia.edu

Abstract

We describe our straight-forward approach for Tasks 5 and 6 of 2021 Social Media Mining for Health Applications (SMM4H) shared tasks. Our system is based on fine-tuning Distill-BERT on each task, as well as first fine-tuning the model on the other task. We explore how much fine-tuning is necessary for accurately classifying tweets as containing self-reported COVID-19 symptoms (Task 5) or whether a tweet related to COVID-19 is self-reporting, non-personal reporting, or a literature/news mention of the virus (Task 6).

1 Introduction

Fine-tuning off-the-shelf Transformer-based contextualized language models is a common baseline for contemporary Natural Language Processing (Ruder, 2021). When developing our system for **Task 6** of the 2021 Social Media Mining for Health Applications (SMM4H), we quickly discovered that fine-tuning DistilBERT (Sanh et al., 2019), a smaller and distilled version of BERT (Devlin et al., 2019), outperformed training traditional, non-neural machine learning models. Fine-tuning DistilBERT on the released training set resulted in a micro-F1 of 97.60 on the Task 6 release development set. While this approach was not as successful for **Task 5** (binary-F1 of 51.49), in this paper, we explore how much fine-tuning is necessary for these tasks and whether there are benefits to first training the model on the other task since both are related to COVID-19.[1]

2 Task Description

Both Task 5 and Task 6 focused on classifying tweets related to COVID-19 (Magge et al., 2021). Task 5 required classifying tweets as describing self-reporting potential cases of COVID-19 or not. Tweets were extracted via manually crafted regular expressions for potential self-reported mentions of COVID-19 and then annotated by two people. $1,148$ Tweets were labeled as containing a self-reporting potential cases and $6,033$ were labeled as "Other." The other tweets that might discuss COVID-19 but do not specifically reporting a user's or their household's potential cases were labeled as "Other."[2] Systems were ranked by F1-score for the "potential case" class.

In Task 6, systems must determine whether a tweet related to COVID-19 is self-reporting, non-personal reporting, or a literature/news mention of the virus. $1,421$ released examples are labeled as self-reporting, $3,567$ as non-personal reports, and $4,464$ as literature/news mentions. Systems were evaluated by micro-F1 score. Table 1 includes examples tweets from the development sets.

3 Method

We fine-tuned DistillBERT using the implementation developed and released by HuggingFace transformer's library (Wolf et al., 2020). We trained the model for 3 epochs, using a batch size of 64 examples, warm-up steps of 500 for the learning rate scheduler and a weight decay of 0.01. Following Peters et al. (2019) recommendation to add minimal task hyper-parameters when fine-tuning pre-trained models, we used the remaining default hyper-parameters from the library's `Trainer` class. All models were trained across 2 NVIDIA RTX 3090's.

3.1 Cross-validation

We used 5-fold evaluation to determine the utility of this simple approach. For each task, we combined the training and development sets and removed duplicate tweets, resulting in $7,174$ and $9,452$ annotated examples for Task 5 and Task 6

[1] All code developed is publicly available at https://github.com/mfleming99/SMM4H_2021.

[2] See Klein et al. (2021) for a detailed description of the data collection and annotation protocols.

Proceedings of the Sixth Social Media Mining for Health Workshop 2021, pages 131–134
June 10, 2021.

Task	Tweet	Label
Task5	Just in case I do manage to contract #coronavirus during the social distancing phase. I will kill it from the INSIDE!	Other
	So I've had this sore throat for a couple of days, I don't know if im being dramatic but i'm scared its Coronavirus??	Potential
Task6	New evidence suggests that neurological symptoms among hospitalized COVID-19 patients are extremely common	Lit-News
	My dad tested positive for COVID-19 earlier this week, started having difficulty breathing this morning, and is now in the ED.	Nonpersonal
	Covid week 13 update. Week 11 kidney pain on the wane, presenting as high BP (affecting brain speed, vision, tightness in veins).	Self Report

Table 1: Examples of tweets and labels for each task, abridged for space.

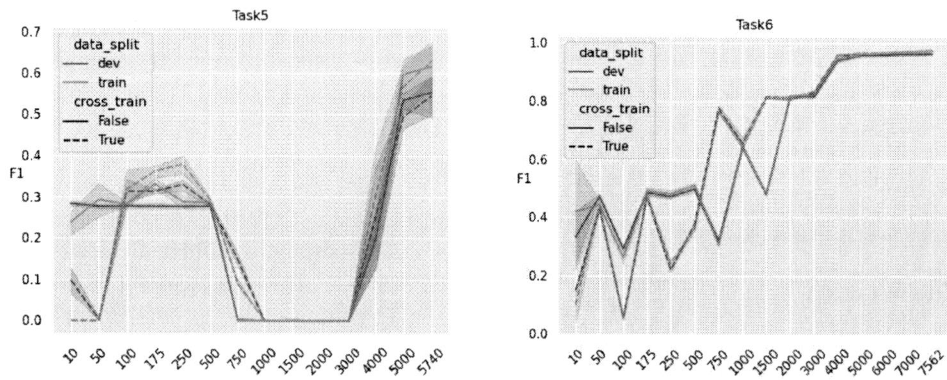

Figure 1: 5-fold results. The left and right graph respectively reflect binary-F1 results for Task 5 and micro-F1 results for Task 6. y-axes indicate F1 and x-axes indicate the number of training examples used. Dotted and solid lines, respectively, indicated that the model was pre-trained on the other task or not. Blue and orange respectively correspond to the training and development folds. The lines indicate the average across the 5 folds and the shaded areas indicate the range of results.

respectively.[3] We divided the datasets into 5 folds of roughly 1,435 and 1,890 labeled examples for Task 5 and Task 6 and fine-tune models on 4 of the folds and test on the held out fold. For each fold, we fine-tuned the model on a increasing number of training examples: 10, 50, 100, 175, 250, 500, 750, 1K, 1.5K, 2K, 3K, 4K, 5K, 6K, 7K, 8K.[4] Additionally, for both tasks, we experimented with using a model pre-trained on the other task. We hypothesized this might be beneficial as these tasks seem to be related.

[3] 7 and 115 examples were removed for Task 5 and 6 respectively.

[4] For Task 5, the maximum number of training examples are 5,740

4 Results

Figure 1 shows the results of fine-tuning DistillBert on each task. For Task 5 (left graph), when fine-tuning on 50 examples or less, initially training on Task 6 (dotted lines) is detrimental. When fine-tuning on somewhere between 50 and 100 training examples, first training the models on Task 6 leads to a noticeable improvement. This continued until we fine-tuned the model on 500 examples. Once we fine-tuned the model on 1000 to 3000 examples, there is no difference between first training on the other task as the models only predict the majority class "Other". As the number of training examples increases from this point, we begin to see large improvements and larger variances between the models trained on different folds. First training on Task 6 appears to be most beneficial when fine-tuning on 100 through 750 Task 5 examples.

For Task 6 (right graph), the benefits of pre-training the model on Task 5 are not as clear cut, and the results oscillate a bit more. It seems that pre-training on Task 5 is only beneficial when fine-tuning the model on 750 through $2,000$ examples (except for the case when fine-tuning on $1,000$ examples). For both tasks, pre-training on the other task seems to make no difference once the model is fine-tuned on enough task specific examples (roughly $1,000$ and $2,000$ examples for Task 5 and Task 6).

Held out test set In these experiments, the model performance on the held out fold seems to increase as we add more training examples. While results for Task 6 seem to plateau, we notice a small increase as we continue to add training examples. Therefore, for our official submissions, we fine-tuned the model on all released examples.

Table 2 reports results for the official test sets.[5] The 63.19 binary-F1 for Task 5 might indicate that training on more examples is beneficial for this task. For Task 6, we notice the micro-F1 drops a bit compared to the results on the held out folds. For both tasks, pre-training on the other task is not beneficial on the test set when trained on as many labeled examples as possible.

We also include a majority vote ensemble of the 5-fold models trained on different training sizes. These test results follow the general trends in Figure 1 indicating when it is most beneficial to first train the DistilBert model on the other task. Similar to the results in Figure 1, when fine-tuning on 750 through $3,000$ Task 5 examples, the model achieved a 0 binary-F1 since it always predicted the majority class "Other."

Train Size	Task5		Task6	
	❄	🔥	❄	🔥
–	63.19	62.24	92.88	91.77
50	29.33	05.72	46.17	42.75
100	–	–	27.65	05.72
175	28.54	32.22	47.97	46.20
250	–	–	46.15	36.83
500	29.29	28.92	-	-
750	00.00	16.00	31.02	56.70
1000	00.00	-	-	-
2000	-	-	80.11	-
4000	-	-	92.41	-
5000	55.69	51.19	-	-

Table 2: Results on the official test sets available on CodaLabs. Numbers indicate binary-F1 for Task 5 and micro-F1 for Task 6. ❄ indicates the model was fine-tuned on the specific task and 🔥 indicates the model was first fine-tuned on the other task. The first line reports the results trained on the combination of the corresponding train and development sets - $7,174$ for Task 5 and $9,452$ for Task 6. The remaining lines are based on a ensemble of the 5 models trained on the corresponding number of examples using a majority vote.

5 Conclusion

We discussed our straightforward approach of fine-tuning a DistilBert model on Tasks 5 and 6 of the 2021 Social Media Mining for Health Applications shared tasks. While not attaining state-of-the-art, these results are competitive and demonstrate the benefit of leveraging large scale pre-trained contextualized language models. We additionally explored the benefits of first training the model on the corresponding task and determine when this can be beneficial. Future work might consider jointly fine-tuning a Bert-based model on both tasks using a multi-task approach as opposed to the transfer learning approach employed here.

[5]These numbers differ from the official leaderboard during the evaluation as we discovered a bug related to loading our pre-trained models during the post-evaluation period.

Acknowledgements

We would like to thank the anonymous reviewer for their feedback. Our experiments were conducted using computational infrastructure provided by the Barnard Vagelos Computational Science Center.

References

Jacob Devlin, Ming-Wei Chang, Kenton Lee, and Kristina Toutanova. 2019. BERT: Pre-training of deep bidirectional transformers for language understanding. In *Proceedings of the 2019 Conference of the North American Chapter of the Association for Computational Linguistics: Human Language Technologies, Volume 1 (Long and Short Papers)*, pages 4171–4186, Minneapolis, Minnesota. Association for Computational Linguistics.

Ari Z Klein, Arjun Magge, Karen O'Connor, Jesus Ivan Flores Amaro, Davy Weissenbacher, and Graciela Gonzalez Hernandez. 2021. Toward using twitter for tracking covid-19: A natural language processing pipeline and exploratory data set. *J Med Internet Res*, 23(1):e25314.

Arjun Magge, Ari Klein, Ivan Flores, Ilseyar Alimova, Mohammed Ali Al-garadi, Antonio Miranda-Escalada, Zulfat Miftahutdinov, Eulàlia Farré-Maduell, Salvador Lima López, Juan M Banda, Karen O'Connor, Abeed Sarker, Elena Tutubalina, Martin Krallinger, Davy Weissenbacher, and Graciela Gonzalez-Hernandez. 2021. Overview of the sixth social media mining for health applications (# smm4h) shared tasks at naacl 2021. In *Proceedings of the Sixth Social Media Mining for Health Applications Workshop & Shared Task*.

Matthew E. Peters, Sebastian Ruder, and Noah A. Smith. 2019. To tune or not to tune? adapting pretrained representations to diverse tasks. In *Proceedings of the 4th Workshop on Representation Learning for NLP (RepL4NLP-2019)*, pages 7–14, Florence, Italy. Association for Computational Linguistics.

Sebastian Ruder. 2021. Recent Advances in Language Model Fine-tuning. `http://ruder.io/recent-advances-lm-fine-tuning`.

Victor Sanh, Lysandre Debut, Julien Chaumond, and Thomas Wolf. 2019. Distilbert, a distilled version of bert: smaller, faster, cheaper and lighter. *arXiv preprint arXiv:1910.01108*.

Thomas Wolf, Lysandre Debut, Victor Sanh, Julien Chaumond, Clement Delangue, Anthony Moi, Pierric Cistac, Tim Rault, Remi Louf, Morgan Funtowicz, Joe Davison, Sam Shleifer, Patrick von Platen, Clara Ma, Yacine Jernite, Julien Plu, Canwen Xu, Teven Le Scao, Sylvain Gugger, Mariama Drame, Quentin Lhoest, and Alexander Rush. 2020. Transformers: State-of-the-art natural language processing. In *Proceedings of the 2020 Conference on Empirical Methods in Natural Language Processing: System Demonstrations*, pages 38–45, Online. Association for Computational Linguistics.

Classification of COVID19 tweets using Machine Learning Approaches

Anupam Mondal[1], Sainik Kumar Mahata[2], Monalisa Dey[3], Dipankar Das[4]
[1,2,3] Institute of Engineering and Management, Kolkata, India
[4] Jadavpur University, Kolkata, India
[1]link.anupam@gmail.com, [2]sainik.mahata@gmail.com
[3]monalisa.dey.21@gmail.com, [4]dipankar.dipnil2005@gmail.com

Abstract

The reported work is a description of our participation in the "Classification of COVID19 tweets containing symptoms" shared task, organized by the "Social Media Mining for Health Applications (SMM4H)" workshop. The literature describes two machine learning approaches that were used to build a three-class classification system, that categorizes tweets related to COVID19, into three classes, viz., self-reports, non-personal reports, and literature/news mentions. The steps for pre-processing tweets, feature extraction, and the development of the machine learning models, are described extensively in the documentation. Both the developed learning models, when evaluated by the organizers, garnered F1 scores of 0.93 and 0.92 respectively.

1 Introduction

In order to identify personal tweets related to COVID-19, it becomes necessary to distinguish them from tweets made by others related to this issue. Classification of medical symptoms from posts related to COVID-19 poses two major challenges: Firstly, the amount of information available as news articles, scientific papers etc that describe various medical symptoms is huge (Mondal et al., 2017; Kushwaha et al., 2020). All this information makes it extremely difficult to spot significant user reported information. Secondly, there are multiple users who report information which is not experienced by themselves but by other people they know or come across (Mondal et al., 2018; Li et al., 2020). This makes the task of identifying self reported information from the huge amount of discourse available very complex.

The current shared task (Task No. 6)[1], namely "Classification of COVID19 tweets containing symptoms" provided participants with three classes viz. **i.** self-reports **ii.** non-personal reports, and **iii.** literature/news mentions. This task is a three way classification task.

For developing the learning models, we used traditional Machine Learning (ML) and state-of-the-art Deep Learning (DL) approaches (Imran et al., 2020; Chakraborty et al., 2020; Gencoglu, 2020). Besides, extra features, like Parts-of-Speech (POS) tags as well as Term Frequency-Inverse Document Frequency (TF-IDF) was used, which enabled the developed models to learn the hidden classes better.

Upon evaluation, our developed models performed well and this was ratified by the fact that they garnered F1 scores of 0.93 (ML model) and 0.92 (DL model) respectively.

The rest of the paper is organized as follows. Section 2 describes the data and methodology that was used to develop both the models. This section describes the pre-processing steps, will talk about the extra features that were used, and will also narrate the learning models that were used to build our systems. Following this, Section 3 and 4 will chronicle the results and the concluding remarks respectively.

2 Methodology

Initially, the organizers provided us with 9,567 training data and 500 validation data. This labeled dataset consisted of three fields; tweet id, the actual tweet, and the respective label (self-reports, non-personal reports, and literature/news mentions). The training and validation data were later combined and it was pre-processed for further development. Steps of pre-processing the tweets included the removal of extra characters to clean the data. The extra characters that were removed/cleaned included mentions, punctuation's and URLs. Additionally, words from hashtags were extracted and extra spaces were contracted. After the pre-processing steps, POS tags of individual words of every tweet were found out using the python pack-

[1]https://healthlanguageprocessing.org/smm4h-2021/task-6/

Proceedings of the Sixth Social Media Mining for Health Workshop 2021, pages 135–137
June 10, 2021.

ages Natural Language Toolkit[2] (NLTK). POS tag features were used as they can help in determining authorship as people's use of words varies. On the other hand, it can easily differentiate between the same words, applied in different settings. E.g., "like" is a verb semantically charged with positive weight, as in "I like you", but it becomes neutral conjunction, as in "I am like you".

For feeding the extra POS tag feature along with individual words, to the ML model, we concatenated them to form an extended input of the structure

$$W_1\_P_1 \quad W_2\_P_2 \quad .... \quad W_n\_P_n$$

where W are the word and P are the POS tag of the word. After the concatenation was done, the input was fed to a TF-IDF vectorizer, which converts a collection of raw documents to a matrix of TF-IDF features. In order to extract the most descriptive terms in a document, we have used TF-IDF features. Besides, this feature assists in computing the similarity between two words which enhances the feature quality to allow even simple models to outperform more advanced ones.

Additionally, the corresponding labels of the tweets were fed to a Label Encoder, which encodes target labels with a value between 0 and n_classes - 1, where n_classes in our case was 3.

Both these vectorized inputs and encoded outputs were fed to a Multi-layer Perceptron classifier (MLP), where alpha was kept at 1 and maximum iterations were kept at 1,000.

For training the DL model, we took the words, POS tags, and TF-IDF values as separate inputs passed them through their respective default Tensorflow embedding layer, and concatenated the outputs. The output, from the concatenation layer, was then passed through two layers of bi-directional Long-Short Term Memory (Hochreiter and Schmidhuber, 1997) (LSTMs) and finally fed to a dense layer which mapped the tensors to the respective labels.

Other parameters of the model were as follows. Optimizer was kept as "adam" and loss was kept as "SparseCategoricalCrossentropy". The batch size was kept as 128 and the number of epochs was fixed at 50. Also, the early stopping mechanism, where the metric was fixed to validation loss, was applied to stop over-fitting. A depiction of the developed model is shown in Figure 1.

[2]https://www.nltk.org/

Both the ML and DL models were then deployed on the 500 validation data provided by the organizers and upon submission, garnered F1 scores of 0.98 and 0.97 respectively. Other validation metrics are shown in Table 1.

Model	Precision	Recall	F1 Score
ML model	0.9720	0.9881	0.98
DL model	0.9660	0.9660	0.97

Table 1: Evaluation scores of the developed models.

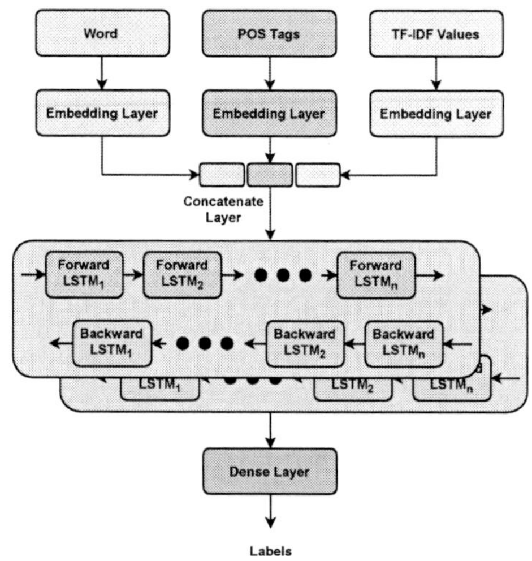

Figure 1: Developed deep learning model for the classification.

3 Evaluation

6,500 tweets were provided by the organizers of the shared task, as test data. Both the developed models were deployed on the same and the results were submitted for evaluation. Upon evaluation, our models garnered micro f1 scores of 0.93 and 0.92, for the ML and DL models respectively. Other scores are shown in Table 2.

Model	Precision	Recall	F1 Score
ML model	0.9337	0.9337	0.93
DL model	0.9248	0.9248	0.92

Table 2: Evaluation scores of the developed models.

Additionally, the arithmetic median of all submissions made by other participating teams is shown in Table 3.

4 Conclusion

The reported system paper presents two models developed using ML and DL approaches, that were trained to classify tweets related to COVID19, into personal/non-personal mentions or standard literature. From the results, we can see that since the amount of training data was low, traditional ML

Precision	Recall	F1 Score
0.93235	0.9337	0.93

Table 3: Median scores of all the participating teams.

methods performed very well. On the contrary, the proposed DL model performed as well, if not better, on the same less amount of data. This is an interesting observation as, more often than not, DL methods rely on huge amounts of data for learning patterns. As future work, we plan to expand this work, by increasing the data and applying state-of-the-art embedding methods like BERT, RoBERTa, etc., on the same.

References

Koyel Chakraborty, Surbhi Bhatia, Siddhartha Bhattacharyya, Jan Platos, Rajib Bag, and Aboul Ella Hassanien. 2020. Sentiment analysis of covid-19 tweets by deep learning classifiers—a study to show how popularity is affecting accuracy in social media. *Applied Soft Computing*, 97:106754.

Oguzhan Gencoglu. 2020. Large scale, language-agnostic discourse classification of tweets during covid-19. *Machine Learning and Knowledge Extraction*, 2(4):603–616.

Sepp Hochreiter and Jürgen Schmidhuber. 1997. Long short-term memory. *Neural computation*, 9(8):1735–1780.

Ali Shariq Imran, Sher Muhammad Daudpota, Zenun Kastrati, and Rakhi Batra. 2020. Cross-cultural polarity and emotion detection using sentiment analysis and deep learning on covid-19 related tweets. *IEEE Access*, 8:181074–181090.

Shashi Kushwaha, Shashi Bahl, Ashok Kumar Bagha, Kulwinder Singh Parmar, Mohd Javaid, Abid Haleem, and Ravi Pratap Singh. 2020. Significant applications of machine learning for covid-19 pandemic. *Journal of Industrial Integration and Management*, 5(4).

Irene Li, Yixin Li, Tianxiao Li, Sergio Alvarez-Napagao, Dario Garcia-Gasulla, and Toyotaro Suzumura. 2020. What are we depressed about when we talk about covid-19: Mental health analysis on tweets using natural language processing. In *International Conference on Innovative Techniques and Applications of Artificial Intelligence*, pages 358–370. Springer.

Anupam Mondal, Erik Cambria, Dipankar Das, and Sivaji Bandyopadhyay. 2017. Employing sentiment-based affinity and gravity scores to identify relations of medical concepts. In *2017 IEEE Symposium Series on Computational Intelligence (SSCI)*, pages 1–7. IEEE.

Anupam Mondal, Dipankar Das, and Sivaji Bandyopadhyay. 2018. A content-based recommendation system for medical concepts: Disease and symptom. In *15th International Conference on Natural Language Processing*, page 120.

Fine-tuning BERT to Classify COVID19 Tweets Containing Symptoms

Rajarshi Roychoudhury
Dept of Computer Science and Engg.
Jadavpur University, India
rroychoudhury2@gmail.com

Sudip Kumar Naskar
Dept of Computer Science and Engg.
Jadavpur University, India
sudip.naskar@gmail.com

Abstract

Twitter provides a source of patient-generated data that has been used in various population health studies. The first step in many of these studies is to identify and capture Twitter messages (tweets) containing medication mentions. Identifying personal mentions of COVID19 symptoms requires distinguishing personal mentions from other mentions such as symptoms reported by others and references to news articles or other sources. In this article, we describe our submission to Task 6 of the Social Media Mining for Health Applications (SMM4H) Shared Task 2021. This task challenged participants to classify tweets where the target classes are - (1) self-reports, (2) non-personal reports, and (3) literature/news mentions. Our system uses a handcrafted preprocessing and word embeddings from BERT encoder model. We achieve. F1 score of 93%.

1 Introduction

The classification of medical symptoms from COVID-19 Twitter posts presents two key issues. Firstly, there is plenty of discourse around news and scientific articles that describe medical symptoms. While this discourse is not related to any user in particular, it enhances the difficulty of identifying valuable user-reported information. Secondly, many users describe symptoms that other people experience, instead of their own, as they are usually caregivers or relatives of people presenting the symptoms. This makes the task of separating what the user is self-reporting particularly tricky, as the discourse is not only around personal experiences. Moreover, detecting tweets containing health-related words such as diseases, treatments and medications is a fundamental yet difficult step. These difficulties are exacerbated by the short length and informal nature of tweets, which often contain non-standard grammar, frequent misspellings, many contractions, extensive slang, and combined symbols (emojis/emoticons) to express emotion (Dang et al., 2020).

From the types of class-labels that are to be predicted, it is clear that contextual representations play an important role beside semantics. Recurrent models are typically used for this task which computes along the symbol positions of the input and output sequences. Aligning the positions to steps in computation time, they generate a sequence of hidden states h_t, as a function of the previous hidden state h_{t-1} and the input for position t. This inherently sequential nature precludes parallelization within training examples, which becomes critical at longer sequence lengths, as memory constraints limit batching across examples. Earlier in context-representation there were two strategies for applying pre-trained language representations to downstream tasks: feature-based and fine-tuning. However, both are limited to the fact that they are unidirectional language models and are unable to learn general language representations. The latest advances in Bidirectional Encoder Representations from Transformers (BERT) address both of these issues, as it is designed to pretrain deep bidirectional representations from unlabeled text by jointly conditioning on both left and right context in all layers (Devlin et al., 2019). We have used small-BERT preprocessing and encoding to get vector representation of sentences, and finetuned BERT for the ternary classification task.

2 Data

Task 6 of SMM4H Shared Task 2021(Magge et al., 2021) challenged participants to develop an automatic classification system to identify tweets mentioning (1) self-reports, (2) non-personal reports, and (3) literature/news mentions. The task was formulated as a multi-class classification task, in which given a set of tweets a system should predict the label for each tweet. Table 1 gives the statistics of the dataset.

Proceedings of the Sixth Social Media Mining for Health Workshop 2021, pages 138–140
June 10, 2021.

Dataset	LN	NP	Self	total
Train	4277	3442	1248	9067
Validation	247	180	73	500
Test	-	-	-	6500

Table 1: Statistics of the dataset. LN: Lit-News, NP: Non-personal

3 System Description

3.1 Data Preprocessing

All apostrophe containing words were expanded. Characters like : , & ! ? were removed . Words were lower-cased to avoid capitalized version of the same word being treated as a different word. The emojis were removed using Python "emoji" library. Hashtags, mentions (words beginning with @) and urls were also removed.

3.2 Model

We used Small_BERT (Tsai et al., 2019) encoder and preprocessing models to extract features from the sentence and used the pooled outputs from the encoder and fed it into a fully connected dense layer, a dropout layer (dropout rate=0.1) and a final dense layer with softmax activation. We used the learning rate of 3e-5 and the adam optimizer. We tested it for 5-10 epochs and obtained the best result after training the model for 9 epochs.

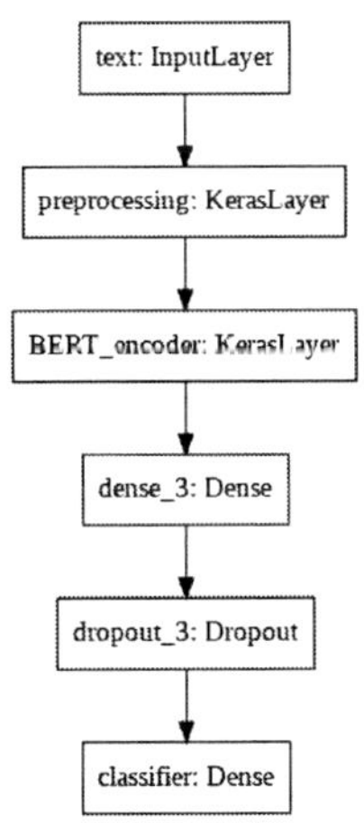

Figure 1: Model

4 Results and Analysis

We obtained an F1 score of 0.968 on the validation set and 0.9325 on the test set (cf. Table 2).

On analysing the wrongly classified tweets in the validation set, we observed some interesting patterns. The sentence "Me and my girl swear we have already had COVID-19. We were sick for nearly a month, fever, cough, sore throat, the doctors told me I had the flu combined with bronchitis because some days I felt like I was drowning in chest mucus." was classified as self-report, while it is an ambiguous case of self-report and non-personal review. The misclasssification of the tweet "I had crippling body aches, fatigue and couldn't concentrate' - was @tomhanksanother COVID19 long-hauler? Sounds v. familiar! LongCovid @HadleyFreeman @guardian" was due to shortcoming of the preprocessing. This tweet is originally labelled as a Lit-News, though it was classified as self report. The main reason is that after preprocesing all the hashtags and the mentions were removed; therefore the overall context the model understood was in first-person and hence it classified the tweet as a self report.

Dataset	F1	Precision	Recall
Validation	.968	.968	.968
Test	.9325	.9325	.9300

Table 2: Results

5 Conclusion

We present a Small BERT based model with custom preprocessing to classify tweets containing COVID-19 symptoms. We tested the model by adjusting various hyperparameters and presented the best result that we obtained using this model. We achieved F1 score of 93%. We observed that the preprocessing needed to include the mentions for some tweets for proper classification, though it was necessary to remove the mentions for the overall increase in the performance of the system.

References

Huong Dang, Kahyun Lee, Sam Henry, and Özlem Uzuner. 2020. Ensemble BERT for classifying medication-mentioning tweets. In *Proceedings of the Fifth Social Media Mining for Health Applications Workshop & Shared Task*, pages 37–41, Barcelona, Spain (Online). Association for Computational Linguistics.

Jacob Devlin, Ming-Wei Chang, Kenton Lee, and Kristina Toutanova. 2019. BERT: Pre-training of deep bidirectional transformers for language understanding. In *Proceedings of the 2019 Conference of the North American Chapter of the Association for Computational Linguistics: Human Language Technologies, Volume 1 (Long and Short Papers)*, pages 4171–4186, Minneapolis, Minnesota. Association for Computational Linguistics.

Arjun Magge, Ari Klein, Ivan Flores, Ilseyar Alimova, Mohammed Ali Al-garadi, Antonio Miranda-Escalada, Zulfat Miftahutdinov, Eulàlia Farré-Maduell, Salvador Lima López, Juan M Banda, Karen O'Connor, Abeed Sarker, Elena Tutubalina, Martin Krallinger, Davy Weissenbacher, and Graciela Gonzalez-Hernandez. 2021. Overview of the sixth social media mining for health applications (# smm4h) shared tasks at naacl 2021. In *Proceedings of the Sixth Social Media Mining for Health Applications Workshop & Shared Task*.

Henry Tsai, Jason Riesa, Melvin Johnson, Naveen Arivazhagan, Xin Li, and Amelia Archer. 2019. Small and practical bert models for sequence labeling. *arXiv preprint arXiv:1909.00100*.

Identifying professions & occupations in Health-related Social Media using Natural Language Processing

J. Alberto Mesa Murgado and **Ana Belén Parras Portillo** and **Pilar López-Úbeda** and **M. Teresa Martín-Valdivia** and **L. Alfonso Ureña López**
SINAI Research Group - CEATIC - Universidad de Jaén
Campus Las Lagunillas s/n. E-23071, Jaén, Spain
{jmurgado,abparras,plubeda,maite,laurena}@ujaen.es

Abstract

This paper describes the entry of the research group SINAI at SMM4H's ProfNER task on identifying professions and occupations in social media data related to health. Specifically, we participated in Task 7a: Tweet Binary Classification to determine whether a tweet contains mentions of occupations or not and also in Task 7b: NER Offset Detection and Classification aimed at predicting occupations mentions and classify them as either professions or working statuses

1 Introduction

Natural Language Processing (NLP) and Machine Learning (ML) techniques are becoming essential in critical fields such as the one of healthcare, considering that they perform tasks faster than a human agent and at a very high level of reliability. Some of these tasks include the automatic assignment of International Classification of Diseases (ICD) codes to health related texts (Perea-Ortega et al., 2020) or the detection of negative and positive emotions in medical documents (Plaza-del Arco et al., 2019).

Automatic text classification and Named Entity Recognition (NER) are two tasks in which NLP has proved to have a relevant impact. In both cases we are given a certain set of documents and while for the first task we aim to classify them distinguishing by a certain criteria, the second seeks to detect and tag specific entities.

The Social Media Mining for Health (SMM4H) 2021 ProfNER Shared Task (Miranda-Escalada et al., 2021) emphasizes the importance of identifying professions and occupations within social media content related to health, this knowledge could later be applied to determine which of them are at risk due to direct exposure to the COVID-19 pandemic and/or state what professional sectors are more prone to mental health issues due to the uncertainty of the current situation.

This issue has been further subdivided into two tracks: determine whether the social media textual content contains mentions of professions or not (a binary classification task) and to identify professions and working statuses within the text in order to extract its text span and tag it accordingly (NER). Our research group has used NLP and ML approaches for both tasks in combination with two dictionaries which we have developed.

Considering this information, this paper is structured as follows: section 2 introduces work related to this challenge and research field. Section 3 briefly describes the dataset provided and its characteristics. Section 4 states the systems we have developed for each task. Section 5 exhibits the results from our systems using the test dataset and finally, in Section 6 we present our conclusions and future work.

2 Related work

Social media plays an important role where people can share information related to health. This information can be used for public health monitoring tasks through the use of NLP techniques.

On one hand, in terms of document classification in the medical field, many researchers have used social networks as a source of information to develop and evaluate systems. For example, to predict mental illnesses such as depression or anorexia (Aldarwish and Ahmad, 2017; López-Úbeda et al., 2021) and to detect nonmedical prescription medication (Al-Garadi et al., 2021). More recently, new studies analyzed health, psychosocial, and social issues emanating from the COVID-19 pandemic from social network comments using NLP (López-Úbeda et al., 2020a; Müller et al., 2020; Oyebode et al., 2020).

On the other hand, several NER systems have been developed using NLP-based systems such as MedLEE (Friedman, 1997), MetaMap (Aronson and Lang, 2010) and cTAKES (Savova et al.,

Proceedings of the Sixth Social Media Mining for Health Workshop 2021, pages 141–145
June 10, 2021.

2010)). Most of these are rule-based systems that use extensive medical vocabularies. Current state-of-the-art approaches to the NER task propose the use of RNNs to learn useful representations automatically because they facilitate the modeling of long-distance dependencies between words in a sentence (López-Úbeda et al., 2019; López-Úbeda et al., 2020b).

Since there is currently a great growth in demand for classification and extraction of information from medical texts, the NLP community has organized a series of open challenges with a focus on biomedical entity extraction and document classification tasks such as DDIExtraction (Segura Bedmar et al., 2013), the N2C2 - National NLP Clinical Challenges shared task (Henry et al., 2020) and the CHEMDNER challenge (Krallinger et al., 2015). Finally, SMM4H (Weissenbacher et al., 2019) provided tasks for the extraction of adverse effects using Twitter as a source of information. In this workshop, participants were first required to identify whether a tweet contained an Adverse Drug Reaction (ADR). Subsequently, the challenge provided the NER task to locate the specific ADR.

The use of Spanish as the main language of a challenge has emerged in recent years providing important workshops such as the DIANN (Fabregat et al., 2018) (Disability Annotation Task) task, PharmaCoNER (Agirre et al., 2019) (Pharmacological Substances, Compounds and proteins and NER), Cantemist (Miranda-Escalada et al., 2020) and eHeatlh-KD (Piad-Morffis et al., 2020) (eHealth knowledge discovery).

3 Dataset

Organizers provided us with a dataset consisting of 8,000 tweets from Twitter subdivided into two subsets: 6,000 tweets with which to train our systems and 2,000 tweets to validate them. Namely, train set and dev set, accordingly.

Besides the tweet's identifier and text span, a binary value was used to determine whether it contained a profession or not as well as its corresponding annotated entities tagged using the Inside-outside-beginning (IBO) format.

To compare the performance of all the systems presented at this shared task, we were provided with another dataset, namely test set, consisting of 2,000 processed tweets and 25,000 raw tweets for the background set.

4 Methodology

For our participation we employ ML models enriched with custom made dictionaries consisting of 776 professions such as *"Farmacéutico"* (Pharmacist), *"Dentista"* (Dentist), *"Cajera"* (Cashier) and *"Veterinario"* (Vet) recovered from the *"Listado de profesiones reguladas en el ámbito sanitario"*[1] provided by the *Ministerio de de Sanidad y Política Social* of the Spanish Government and from the occupations listed by the European Commission in their International Standard Classification of Occupations (ISCO)[2].

The second dictionary contains 26 working statuses including *"Autónomo"*, *"Funcionario"* (Public employee) and *"Erte"* (record of temporary Employment regulation) among others based on the Workshop's annotation guidelines[3].

4.1 Task 7a: Binary Classification

To classify tweets whether if they contained mentions of professions or not, we used two approaches: a Support Vector Machine (SVM) and bidirectional Long Short Term Memory (BiLSTM) Recurrent Neural Network (RNN), both combined with our professions dictionary.

The SVM approach applies the scikit-learn library (Pedregosa et al., 2011) using its default parameters, as stated in the documentation, the words for each tweet, also stated as document, were transformed into vectors considering the frequency of the terms within each document (TFIDF). This structure was later enriched using our professions dictionary through a vector for each document consisting of as much binary values as the number of terms in the dictionary such as each binary value represented if the term within the document was in the dictionary or not.

The BiLSTM model is implemented using the Tensorflow library (Abadi et al., 2015) which we explored through different batch sizes, ranging from 2^7 (2^6 and 2^8), and different number of epochs. Although the best accuracy obtained for training (0.8695) determined a batch size of 128 and 5 epochs. This model makes use of the GloVe (Pennington et al., 2014) word embeddings 200d vector,

[1] https://www.mscbs.gob.es/eu/profesionales/formacion/docs/Anexo_X_del_Real_Decreto_1837.pdf

[2] https://ec.europa.eu/esco/portal/occupation

[3] https://zenodo.org/record/4306017#.YE3vpJ1KhhE

pre-trained using Twitter's tweets, this approach is also uses our professions dictionary.

4.2 Task 7b: NER offset detection and classification

To detect and classify entities within those same tweets and retrieve them discerning the text span of the entities as well as their initial and end position within the text, we opted for a Conditional Random Fields (CRF) approach.

This approach was implemented using scikit-learn crfsuite library (Wijffels and Okazaki, 2007-2018), transforming words to features and using the words that came before and after each term, then enriched the system using both our dictionaries (added features as binary values). For this implementation we considered the L-BFGS method for the gradient descent, a 100 iterations and values 0.1 for both c1 and c2.

Our system assigned an IBO tag to each term of every tweet and we later searched for the entity text span within the same tweet to extract its initial and end position.

5 Evaluation results

We have been provided with the median of the participants' score using three metrics: precision (P), recall (R) and F1-scoring (F1) (Magge et al., 2021). Using them we built 2 tables consisting of two sections: the upper section shows the score obtained by our systems on the test set, while the latter fits the same purpose for the dev set.

5.1 Task 7a: Binary Classification

We submitted 2 runs for the evaluation phase: a combination of SVM (SVM+Dic) and an BiLSTM model (BiLSTM+Dic), both combined with our professions dictionary, the scoring associated to these systems applied to the provided datasets is displayed in Table 1.

Model	P	R	F1
`Median`	**0.9185**	**0.8553**	**0.85**
`BiLSTM+Dic`	0.7612	0.4752	0.59
`SVM+Dic`	0.8995	0.4255	0.58
`BiLSTM+Dic`	0.77	**0.72**	**0.74**
`SVM+Dic`	**0.86**	0.70	**0.74**
`SVM`	**0.86**	0.64	0.66

Table 1: Scores obtained by our systems on the SMM4H ProfNER Shared Task - Task 7a (binary classification) applied over the test and dev set, accordingly.

While the performance of our systems on the training dataset was, at average, close to 0.74 (F1), on the test set it was decreased in a 21%. Therefore, resulting in a value close to 0,59 (again, F1), 31% below the median.

5.2 Task 7b: NER Offset Classification

We were closer to the median and in line with what our systems obtained on the training dataset (1% decrease in performance compared to the test set). These results are displayed in Table 2 in the same way in which Table 1 was: the upper section refers to the scoring for our systems on the test set while the latter, exhibits the scoring for the dev set.

Model	P	R	F1
`Median`	**0.842**	**0.7265**	**0.7605**
`CRF+Dic`	0.824	0.652	0.728
`CRF+Dic`	**0.861**	**0.647**	**0.739**
`CRF`	0.852	0.597	0.702

Table 2: Score obtained by our systems on the SMM4H ProfNER Shared Task - Task 7b (NER) applied over the test and dev set, accordingly.

6 Conclusion

For our participation in the SMM4H Task 7 ProfNER Shared Task on identifying professions and occupations we implemented three systems. First two are aimed at the binary classification task (Task A) using an SVM and a BiLSTM approach, both combined with our professions dictionary. The latter system follows a CRF approach combined with our professions and working statuses dictionaries and it is applied to the NER task (Task B).

Our predictions for the training set were consistent with those obtained on the test set for the second task (NER) whereas our approaches for the first task (binary classification) fell short of our expectations by 21% below our training results.

For future work, we will use the gold test in order to perform a deeper analysis to assess why did this event happened and therefore improve the performance of our systems.

Acknowledgements

This work has been partially supported by the LIVING-LANG project [RTI2018-094653-B-C21] of the Spanish Government and the Fondo Europeo de Desarrollo Regional (FEDER).

References

Martín Abadi, Ashish Agarwal, Paul Barham, Eugene Brevdo, Zhifeng Chen, Craig Citro, Greg S. Corrado, Andy Davis, Jeffrey Dean, Matthieu Devin, Sanjay Ghemawat, Ian Goodfellow, Andrew Harp, Geoffrey Irving, Michael Isard, Yangqing Jia, Rafal Jozefowicz, Lukasz Kaiser, Manjunath Kudlur, Josh Levenberg, Dan Mané, Rajat Monga, Sherry Moore, Derek Murray, Chris Olah, Mike Schuster, Jonathon Shlens, Benoit Steiner, Ilya Sutskever, Kunal Talwar, Paul Tucker, Vincent Vanhoucke, Vijay Vasudevan, Fernanda Viégas, Oriol Vinyals, Pete Warden, Martin Wattenberg, Martin Wicke, Yuan Yu, and Xiaoqiang Zheng. 2015. TensorFlow: Large-scale machine learning on heterogeneous systems. Software available from tensorflow.org.

Aitor Gonzalez Agirre, Montserrat Marimon, Ander Intxaurrondo, Obdulia Rabal, Marta Villegas, and Martin Krallinger. 2019. Pharmaconer: Pharmacological substances, compounds and proteins named entity recognition track. In *Proceedings of The 5th Workshop on BioNLP Open Shared Tasks*, pages 1–10.

Mohammed Ali Al-Garadi, Yuan-Chi Yang, Haitao Cai, Yucheng Ruan, Karen O'Connor, Gonzalez-Hernandez Graciela, Jeanmarie Perrone, and Abeed Sarker. 2021. Text classification models for the automatic detection of nonmedical prescription medication use from social media. *BMC medical informatics and decision making*, 21(1):1–13.

M. M. Aldarwish and H. F. Ahmad. 2017. Predicting depression levels using social media posts. In *2017 IEEE 13th International Symposium on Autonomous Decentralized System (ISADS)*, pages 277–280.

Alan R Aronson and François-Michel Lang. 2010. An overview of metamap: historical perspective and recent advances. *Journal of the American Medical Informatics Association*, 17(3):229–236.

Hermenegildo Fabregat, Juan Martinez-Romo, and Lourdes Araujo. 2018. Overview of the DIANN Task: Disability Annotation Task. In *IberEval@ SEPLN*, pages 1–14.

Carol Friedman. 1997. Towards a comprehensive medical language processing system: methods and issues. In *Proceedings of the AMIA annual fall symposium*, page 595. American Medical Informatics Association.

Sam Henry, Kevin Buchan, Michele Filannino, Amber Stubbs, and Ozlem Uzuner. 2020. 2018 n2c2 shared task on adverse drug events and medication extraction in electronic health records. *Journal of the American Medical Informatics Association*, 27(1):3–12.

Martin Krallinger, Florian Leitner, Obdulia Rabal, Miguel Vazquez, Julen Oyarzabal, and Alfonso Valencia. 2015. Chemdner: The drugs and chemical names extraction challenge. *Journal of cheminformatics*, 7(1):S1.

Pilar López-Úbeda, Manuel Carlos Díaz-Galiano, Teodoro Martín-Noguerol, Antonio Luna, L Alfonso Ureña-López, and M Teresa Martín-Valdivia. 2020a. Covid-19 detection in radiological text reports integrating entity recognition. *Computers in Biology and Medicine*, 127:104066.

Pilar López-Úbeda, Manuel Carlos Díaz Galiano, M Teresa Martín-Valdivia, and L Alfonso Urena Lopez. 2019. Using machine learning and deep learning methods to find mentions of adverse drug reactions in social media. In *Proceedings of the Fourth Social Media Mining for Health Applications (# SMM4H) Workshop & Shared Task*, pages 102–106.

Pilar López-Úbeda, José M Perea-Ortega, Manuel C Díaz-Galiano, M Teresa Martín-Valdivia, and L Alfonso Ureña-López. 2020b. Sinai at ehealth-kd challenge 2020: Combining word embeddings for named entity recognition in spanish medical records.

Pilar López-Úbeda, Flor Miriam Plaza-del Arco, Manuel Carlos Díaz-Galiano, and Maria-Teresa Martín-Valdivia. 2021. How successful is transfer learning for detecting anorexia on social media? *Applied Sciences*, 11(4):1838.

Arjun Magge, Ari Klein, Ivan Flores, Ilseyar Alimova, Mohammed Ali Al-garadi, Antonio Miranda-Escalada, Zulfat Miftahutdinov, Eulàlia Farré-Maduell, Salvador Lima López, Juan M Banda, Karen O'Connor, Abeed Sarker, Elena Tutubalina, Martin Krallinger, Davy Weissenbacher, and Graciela Gonzalez-Hernandez. 2021. Overview of the sixth social media mining for health applications (# smm4h) shared tasks at naacl 2021. In *Proceedings of the Sixth Social Media Mining for Health Applications Workshop & Shared Task*.

A Miranda-Escalada, E Farré, and M Krallinger. 2020. Named entity recognition, concept normalization and clinical coding: Overview of the cantemist track for cancer text mining in spanish, corpus, guidelines, methods and results. In *Proceedings of the Iberian Languages Evaluation Forum (IberLEF 2020), CEUR Workshop Proceedings*.

Antonio Miranda-Escalada, Eulàlia Farré-Maduell, Salvador Lima López, Vicent Briva-Iglesias, Marvin Agüero-Torales, Luis Gascó-Sánchez, and Martin Krallinger. 2021. The profner shared task on automatic recognition of professions and occupation mentions in social media: systems, evaluation, guidelines, embeddings and corpora. In *Proceedings of the Sixth Social Media Mining for Health Applications Workshop & Shared Task*.

Martin Müller, Marcel Salathé, and Per E Kummervold. 2020. Covid-twitter-bert: A natural language processing model to analyse covid-19 content on twitter. *arXiv preprint arXiv:2005.07503*.

Oladapo Oyebode, Chinenye Ndulue, Ashfaq Adib, Dinesh Mulchandani, Banuchitra Suruliraj, Fidelia Anulika Orji, Christine Chambers, Sandra Meier, and Rita Orji. 2020. Health, psychosocial, and social issues emanating from covid-19 pandemic based on social media comments using natural language processing. *arXiv preprint arXiv:2007.12144*.

F. Pedregosa, G. Varoquaux, A. Gramfort, V. Michel, B. Thirion, O. Grisel, M. Blondel, P. Prettenhofer, R. Weiss, V. Dubourg, J. Vanderplas, A. Passos, D. Cournapeau, M. Brucher, M. Perrot, and E. Duchesnay. 2011. Scikit-learn: Machine learning in Python. *Journal of Machine Learning Research*, 12:2825–2830.

Jeffrey Pennington, Richard Socher, and Christopher D. Manning. 2014. Glove: Global vectors for word representation. In *Empirical Methods in Natural Language Processing (EMNLP)*, pages 1532–1543.

José M Perea-Ortega, Pilar López-Úbeda, Manuel C Díaz-Galiano, M Teresa Martín-Valdivia, and L Alfonso Ureña-López. 2020. Sinai at clef ehealth 2020: testing different pre-trained word embeddings for clinical coding in spanish.

Alejandro Piad-Morffis, Yoan Gutiérrez, Hian Cañizares-Diaz, Suilan Estevez-Velarde, Rafael Muñoz, Andres Montoyo, Yudivian Almeida-Cruz, et al. 2020. Overview of the ehealth knowledge discovery challenge at iberlef 2020. CEUR.

Flor Miriam Plaza-del Arco, M Dolores Molina-González, M Teresa Martín-Valdivia, and L Alfonso Ureña-López. 2019. Sinai at semeval-2019 task 3: Using affective features for emotion classification in textual conversations. In *Proceedings of the 13th International Workshop on Semantic Evaluation*, pages 307–311.

Guergana K Savova, James J Masanz, Philip V Ogren, Jiaping Zheng, Sunghwan Sohn, Karin C Kipper-Schuler, and Christopher G Chute. 2010. Mayo clinical text analysis and knowledge extraction system (ctakes): architecture, component evaluation and applications. *Journal of the American Medical Informatics Association*, 17(5):507–513.

Isabel Segura Bedmar, Paloma Martínez, and María Herrero Zazo. 2013. Semeval-2013 task 9: Extraction of drug-drug interactions from biomedical texts (ddiextraction 2013). Association for Computational Linguistics.

Davy Weissenbacher, Abeed Sarker, Arjun Magge, Ashlynn Daughton, Karen O'Connor, Michael Paul, and Graciela Gonzalez. 2019. Overview of the fourth social media mining for health (smm4h) shared tasks at acl 2019. In *Proceedings of the fourth social media mining for health applications (# SMM4H) workshop & shared task*, pages 21–30.

Jan Wijffels and Naoaki Okazaki. 2007-2018. crfsuite: Conditional random fields for labelling sequential data in natural language processing based on crfsuite: a fast implementation of conditional random fields (crfs). R package version 0.1.

Approaching SMM4H with auto-regressive language models and back-translation

Joseph Cornelius[*†] **Tilia Ellendorff**[‡†] **Fabio Rinaldi**[*†]
[*]Dalle Molle Institute for Artificial Intelligence Research (IDSIA)
[†]Swiss Institute of Bioinformatics
[‡]University of Zurich, Department of Computational Linguistics
{joseph.cornelius,fabio.rinaldi}@idsia.ch
tilia.ellendorff@uzh.ch

Abstract

We describe our submissions to the 6th edition of the Social Media Mining for Health Applications (SMM4H) shared task. Our team (OGNLP) participated in the sub-task: Classification of tweets self-reporting potential cases of COVID-19 (Task 5). For our submissions, we employed systems based on auto-regressive transformer models (XLNet) and back-translation for balancing the dataset.

1 Introduction

The Social Media Mining for Health Applications (SMM4H) shared task 2021 (Magge et al., 2021) focuses on textual processing of noisy social media data in the health domain. Our team (OGNLP) participated in Task 5, a binary classification task to identify tweets self-reporting potential cases of COVID-19. Tweets are labeled as positive (marked "1") if they self-report potential cases of COVID-19, and negative (marked "0") otherwise.

2 Dataset

The data provided by the organizers comprises tweets gathered from Twitter. As shown in Table 1, we have a total of 6465 training samples plus additional 716 validation samples. A test set with 10000 samples is provided for evaluation. On average, each tweet has 38 tokens and 155 characters. The dataset is unbalanced since the amount of negatively labeled tweets is five times higher than that of positively labeled tweets.

3 Methods

3.1 Preprocessing

Textual data from social media is often noisy since it contains many misspellings, abbreviations, emoticons, and non-standard wordings. Thus, preprocessing is a crucial part to de-noising the dataset and therefore increasing the performance. For this purpose, we modified the tweets as follows:

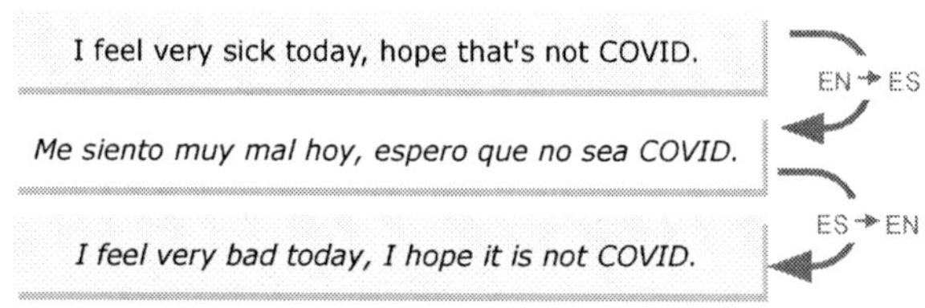

Figure 1: An example of back-translation.

- Hash symbols (#) were stripped from hash tags.
- All punctuation characters except ".,!?" were removed.
- URLs were eliminated from tweets.
- Emojies were stripped from tweets.
- The lowercase version of all tweets was used.

3.2 XLNet

As a baseline model, we used a pre-trained transformer language model, XLNet, which achieves state-of-the-art results on several sentiment analysis datasets (Yang et al., 2019). In contrast to the popular transformer-based BERT model (Devlin et al., 2019), XLNet is not pre-trained by predicting a masked token solely conditioned on its left (or right) tokens within the sentence, but instead the objective is to predict a masked token conditioned on all permutations of tokens within the sentence. Thus, XLNet's distinguishing feature is that it is able to learn context bidirectionally by permuting all the words in a sentence.

For the different trials, we used "XLNet-large-cased" from Huggingfaces python API[1] with a consistent setup of hyperparameters. We truncated each tweet to a maximum length of 256 characters, applied a batch size of four, and used a learning rate of 3e-6.

[1]https://huggingface.co/transformers/model_doc/xlnet.html

Proceedings of the Sixth Social Media Mining for Health Workshop 2021, pages 146–148
June 10, 2021.

Dataset	Neg	Pos	Total
Train	5,439	1,026	6,465
Valid	594	122	716
Test	-	-	10,000

Table 1: The number of tweets provided for Task 5, divided into training, validation, and test datasets.

System	Precision	Recall	F-Score
Validation Set			
XLNet	0.83	0.81	0.82
XLNet+BT$_{k=1}$	**0.83**	**0.86**	**0.84**
XLNet+BT$_{k=1}$+PM	0.79	0.80	0.79
Test Set			
XLNet	0.66	0.72	0.69
XLNet+BT$_{k=1}$	**0.70**	**0.72**	**0.72**
Mean	*0.74*	*0.74*	*0.75*

Table 2: Official and unofficial results of our systems, compared to the mean score of all competing systems.

3.3 Back-translation

Back-translation (BT) is a form of data argumentation and takes advantage of the advances in machine translation (Sennrich et al., 2015). BT allows us to balance the training set through the increase of the number of positive samples. Here our goal is to obtain a paraphrased tweet t' of a tweet t. To this end we automatically translate t into a different language (pivot) yielding $\tilde{t}$. Subsequently, we translate $\tilde{t}$ back to the source language and thus obtain the paraphrased tweet t'. BT leverages the fact that a translation often has several equivalent expressions in the target language. To obtain the BT dataset D_{BT}, we used the Google Translation API through TextBlob[2] and back-translated each English tweet from the minority class using the following ten languages as pivot: Bulgarian, Dutch, Gujarati, Hindi, Igbo, Japanese, Maltese, Pashto, Persian and Spanish. To increase the variance of the BT, we included pivots from low-resource languages and different language families.

Figure 1 shows that we can retrieve the paraphrased tweet t' *"I feel very bad today, I hope it is not COVID. "* from the original tweet t *"I feel very sick today, hope that's not COVID."* by using a BT from English → Spanish → English.

3.4 Parameter Merging

Parameter merging (PM) of equivalent models trained on different subsets of a dataset can be used

[2] `https://textblob.readthedocs.io`

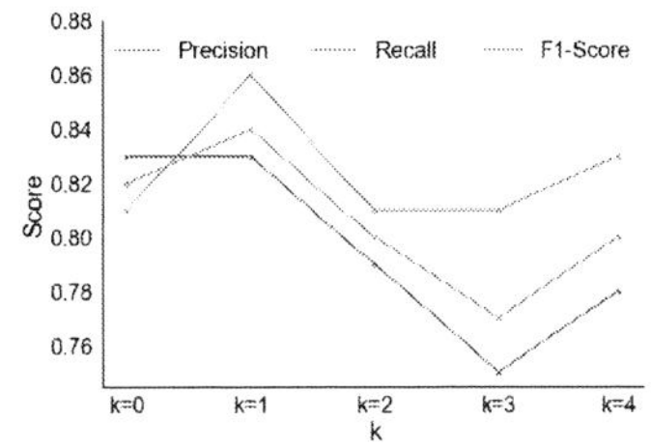

Figure 2: Influence of the number of BT samples used per tweet (k) on the performance of the XLNet system.

to obtain a more robust and more generalized model (Utans, 1996; Ellendorff et al., 2019). For this purpose, we created a merged XLNet system from five XLNet models obtained by five-fold stratified cross-validation. For the merged XLNet system, we calculated the parameters' average across all five XLNet models.

4 Results and Discussion

Table 2 shows the official results on the test set, as well as the unofficial results on the validation set. For models incorporating BT, the number of BT samples randomly drawn from D_{BT} used for each tweet is given by k, which means that for k=2, we triple the number of training samples. The XLNet model with BT k=1 has achieved the best results on both the test and validation dataset with an F-score of 0.72 and 0.84 respectively. In Figure 2 we can see that, contrary to expectation, the F-score of the models trained with BT does not constantly increase with an increase in k and has its optimum at k=1. The XLNet system generated from the PM of 5 XLNet models trained with cross-validation and BT k=1 achieved only the third-best F-score 0.80 on the validation dataset. Hence, we did not select the PM system for official submission as the submission was limited to the results of two runs. For the second official submission, we used the XLNet system trained for four epochs without BT. However, this model achieved a significantly lower F-score with 0.69 on the official test set and 0.82 on the unofficial test set.

We assume that PM did not lead to an improvement as we had to set the number of folds for cross-validation very low (5) due to the limited GPU computing power available to us. Furthermore, we can conclude that back-translation leads to a significant improvement, but the number of additional generated samples plays a decisive role.

References

Jacob Devlin, Ming-Wei Chang, Kenton Lee, and Kristina Toutanova. 2019. BERT: pre-training of deep bidirectional transformers for language understanding. In *Proceedings of the 2019 Conference of the North American Chapter of the Association for Computational Linguistics*, pages 4171–4186. Association for Computational Linguistics.

Tilia Ellendorff, Lenz Furrer, Nicola Colic, Noëmi Aepli, and Fabio Rinaldi. 2019. Approaching smm4h with merged models and multi-task learning. In *Proceedings of the Fourth Social Media Mining for Health Applications (# SMM4H) Workshop & Shared Task*, pages 58–61. University of Zurich.

Arjun Magge, Ari Klein, Ivan Flores, Ilseyar Alimova, Mohammed Ali Al-garadi, Antonio Miranda-Escalada, Zulfat Miftahutdinov, Eulàlia Farré-Maduell, Salvador Lima López, Juan M Banda, Karen O'Connor, Abeed Sarker, Elena Tutubalina, Martin Krallinger, Davy Weissenbacher, and Graciela Gonzalez-Hernandez. 2021. Overview of the sixth social media mining for health applications (# smm4h) shared tasks at naacl 2021. In *Proceedings of the Sixth Social Media Mining for Health Applications Workshop & Shared Task*.

Rico Sennrich, Barry Haddow, and Alexandra Birch. 2015. Improving neural machine translation models with monolingual data. *arXiv preprint arXiv:1511.06709*.

Joachim Utans. 1996. Weight averaging for neural networks and local resampling schemes. In *Proc. AAAI-96 Workshop on Integrating Multiple Learned Models. AAAI Press*, pages 133–138. Citeseer.

Zhilin Yang, Zihang Dai, Yiming Yang, Jaime Carbonell, Russ R Salakhutdinov, and Quoc V Le. 2019. Xlnet: Generalized autoregressive pretraining for language understanding. In *Advances in Neural Information Processing Systems*, volume 32. Curran Associates, Inc.

ULD-NUIG at Social Media Mining for Health Applications (#SMM4H) Shared Task 2021

Atul Kr. Ojha, Priya Rani, Koustava Goswami,
Bharathi Raja Chakravarthi, John P. McCrae
Data Science Institute, National University of Ireland Galway
(atulkumar.ojha, priya.rani, koustava.goswami,
bharathi.raja, john.mccrae)@insight-centre.org

Abstract

In this paper, we present the ULD-NUIG team's system, designed as part of Social Media Mining for Health Applications (#SMM4H) Shared Task 2021. We participate in two tasks out of eight, namely "Classification of tweets self-reporting potential cases of COVID-19" (Task 5) and "Classification of COVID19 tweets containing symptoms" (Task 6). The team conduct a series of experiments to explore the challenges of both the tasks. We used a multilingual pre-trained BERT model for Task 5 and Generative Morphemes with Attention (GenMA) model for Task 6. In the experiments, we find that, GenMA, developed for Task 6, gives better results on both validation and test data-set. The submitted systems achieve F-1 score 0.53 for Task 5 and 0.84 for Task 6 on test data-set.

1 Introduction

In recent decades, social media has proved to be one of the greatest sources of information exchange. When the world was overtaken by the COVID-19 outbreak, social media became the greatest platform for the general public to exchange different information about the pandemic. With the widespread digitization of behavioural and medical data, the emergence of social media, and the Internet's infrastructure of large-scale knowledge storage and distribution, there has been a breakthrough in our ability to identify human social interactions, behavioural patterns, cognitive processes and their relationships with healthcare. At the same time, it has also induced a different level of challenges in the natural language processing field such as detection of medical jargons, named entity recognition, multi-word expressions. Furthermore, the informal nature of tweets and short length, which often contain non-standard grammar, frequent misspellings, many contractions, extensive slang, and use of emojis/emoticons to express emotion exacerbate the challenges (Dang et al., 2020). The scenario becomes even more complicated since social media covers very large populations and the geographical location.

Despite several issues, social media data has been used to monitor human health and disease (the recent pandemic outburst) all over the world. Many promising methodologies are being developed. In this paper, we describe two different systems trained on the data provided by the Social Media Mining for Health Applications (#SMM4H) Shared Task 2021 organisers (Magge et al., 2021) namely Task 5: Classification of tweets self-reporting potential cases of COVID-19 and Task 6: Classification of COVID19 tweets containing symptoms (see Section 2). We conduct a series of experiments to explore the challenges of both the tasks. We use multilingual pre-trained BERT model for Task 5 and Generative Morphemes with Attention (GenMA) model for Task 6 (see Section 3).

2 Data Augmentation

2.1 Data Size

Task 5 : Classification of tweets self-reporting potential cases of COVID-19
We use the data set provided by the organisers of Task 5 SMM4H'21.[1] The data was divided into training set, validation set and test set as detailed in Table 1. The task involves binary classification of the tweets, which distinguishes self-reported potential cases of COVID-19 annotated as 1 and non potential cases annotated as 0.

Task 6 : Classification of COVID19 tweets containing symptoms
The data set for the experiment was given by the organisers of Task 6 SMM4H'21.[2] Like Task 5 this data was also divided in training,

[1] https://healthlanguageprocessing.org/smm4h-2021/task-5/

[2] https://healthlanguageprocessing.org/smm4h-2021/task-6/

Proceedings of the Sixth Social Media Mining for Health Workshop 2021, pages 149–152
June 10, 2021.

validation and test set. The statistics of the data set is given in Table 1. The data is classified at three different levels: self-reports, non-personal reports, literature/news mentions.

Task	Training	Validation	Test
Task 5	6,465	717	10,000
Task 6	9,068	501	6,500

Table 1: Statistics of Task 5 and 6 Dataset

2.2 Pre-processing

We normalized the data through the following pre-processing steps as part of the experiment.

1. After a thorough manual evaluation of the data set, we came to the conclusion that emoticons, URLs along with the other special characters which are very common in social media data do not serve necessary purpose for our tasks. Therefore we removed emoticon, URLs and other special characters from the data set.
2. Lower casing all the tweets in the data set. After lower casing the tweets all extra spaces were removed from it.

3 Experiments

3.1 Task 5

We have used the multilingual pre-trained BERT (Devlin et al., 2019; Turc et al., 2019) model to fine-tune our model on the given Task 5 training data set. The detailed model descriptions is given below:

- The model has an embedding dimension of 768. We have used the Google-provided cased vocabulary.
- Parameters for training - We have trained our model for 3 epochs on the training data set and have used a stepped LR scheduler for the learning rate scheduling. The learning rate is set to 2e-5 (see Equation 1).

Based on Hugging Face implementation, we have used the below equation as warmup steps definition for training the model. Here 'r' is the tuneable parameter, which defines the percentage of data used to define the step size while training. We have used 10% of the data while training. After training the model we have tested it on the held-out test data set given by the organizers.

$$W_{steps} = \frac{(len(training_{set}) * epochs_{training})}{batchsize_{training} * r} \quad (1)$$

3.2 Task 6

We have taken the inspiration from the Generative Morphemes with Attention (GenMA) model (Goswami et al., 2020) to develop the model for Task 6. We have noted the model description below:

- The model takes the character sequence as the input sequence. It has one character embedding layer and two convolutions (CONV1D) layers. Each convolution layer has one max-pooling layer. After the convolution layers, there is one LSTM layer and one bidirectional LSTM layer, followed by two self-attention layers. The model has two hidden layers and one softmax layer. The model generates new artificial morphemes and frames a sentence as a group of new morphemes. The combination of two CNN layers helps to generate new morphemes based on deep relative co-occurring characters (3 characters frame), and the LSTM layers help to capture the global information of sentences based on newly generated features. The self-attention layers help to construct sentence-level information. It also captures relativity strength among different co-occurring character features.
- We have used 32 filters, each with a kernel size of 3. The max-pooling size is 3. The hidden size h_i of LSTM units is kept to 100. The dense layer has 32 neurons, and it has 50 percent dropout. The Adam optimizer (Kingma and Ba, 2015) is used to train our model with the default learning set to 0.0001. The batch size is set to 10. For the convolution layer in both the experiments we have used the relu activation function (Nair and Hinton, 2010) and for the dense layer we have used tanh activation function (Kalman and Kwasny, 1992). Categorical cross-entropy loss is used for the multi-class classification. We have used Keras[3] to train and test our model.

[3]https://keras.io

The convolution layers act as the feature extractor of the sentences. The one-dimensional convolution implements one-dimensional filters which slide over the sentences as a feature extractor. The second convolution layer will take feature representations as input and generate a high-order feature representation of the characters. The max-pooling network after each convolution network helps to capture the most important features of size *d*. The new high-order representations are then fed to the LSTM (Long Short Term Memory Network) as input.

The LSTM layer takes the output of the previous CNN layer as input. The LSTM layer produces a new representation sequences in the form of $h_1, h_2, ...h_n$ where h_t is the hidden state of the LSTM of time step *t*, summarising all the information of the input features (morphemes) of the sentences. At each time step, *t*, the hidden state takes the previous time step hidden state h_{t-1} and characters (x_t) as input. A bidirectional LSTM (BiLSTM) network has been used, which has helped us to summarise the information of the features from both directions. The Bidirectional LSTM consists of a forward pass and a backward pass which provides two annotations of the hidden state h_{for} and h_{back}. We obtained the final hidden state representation by concatenating both hidden states $h_i = h_{i-for} \oplus h_{i-back}$, where h_i is the hidden state of the *i*-th timestep and $\oplus$ is the element-wise sum between the matrices.

The attention layer helps to determine the importance of one morpheme over others while building sentence embedding for classification. The self-attention mechanism has been adopted from Baziotis et al. (2018), which helped to identify the morphemes that capture the important features to classify the tweets. The mechanism assigns weight a_i to each feature's annotation based on output h_i of the BiLSTM's hidden states, with the help of the softmax function as illustrated in Equation 2 and 3 (Baziotis et al., 2018).

$$a_i = \tanh(W_h \cdot h_i + b_h) \tag{2}$$

$$a_i = \frac{exp(a_i)}{\sum_{t=1}^{T} exp(a_t))} \tag{3}$$

The new representation will give a fixed representation of the sentence by taking the weighted sum of all feature-label annotations as shown in Equation 4.

$$r = \sum_{i=1}^{T} a_i \cdot h_i \tag{4}$$

where W_h and b_h are the attention weights and bias respectively (Baziotis et al., 2018; Goswami et al., 2020).

The output layer consists of one fully-connected layer with one softmax layer. The sentence representation after the attention layer is the input for the dense layer. The output of the dense layer is the input of the softmax which gives the probability distribution of all the classes with the help of the softmax function.

4 Evaluation

We use shared task organizers' validation and test data-set for evaluation. The standard evaluation metrics, Precision, Recall and F-1 score, were used for automatic evaluation. It gives a quantitative picture of particular differences across different systems, especially with reference to evaluation scores. On the validation data-set, Task 5 and 6 systems' F-1 score were 0.89 and 0.95 respectively. While on the test data-set, F-1 score were 0.53 and 0.84 respectively. The detailed results are given in Table 2.

System	Precision	Recall	F-1 score
Task 5	0.7412	0.4091	0.53
Task 6	0.8415	0.8415	0.84

Table 2: Accuracy of Task-5 and 6 Systems on Test Data-set

5 Summing up

The entire series of experiments gave us various types of insights to deal with social media data for mining medical information. We observed that pre-trained language models such as BERT do not provide good results for extraction of medical information for COVID-19. One of the reasons that we could think is that these models are trained on various domain data set but it is very unlikely that these data-sets contain information regarding COVID-19. On the other hand our characters based attention model outperform the BERT model. In future, we would like to explore more models with word features, specific linguistic features in order to deeply understand the characteristics of social media mining for clinical information.

Acknowledgements

This publication has emanated from research in part supported by the EU H2020 programme under grant agreements 731015 (ELEXIS-European Lexical Infrastructure) as well as by the Irish Research Council under grant number SFI/18/CRT/6223 (CRT-Centre for Research Training in Artificial Intelligence) co-funded by the European Regional Development Fund.
We are also grateful to the organizers of SMM4H'21 Shared Task for providing us the Task 5 and 6 data and evaluation scores.

References

Christos Baziotis, Athanasiou Nikolaos, Pinelopi Papalampidi, Athanasia Kolovou, Georgios Paraskevopoulos, Nikolaos Ellinas, and Alexandros Potamianos. 2018. NTUA-SLP at SemEval-2018 task 3: Tracking ironic tweets using ensembles of word and character level attentive RNNs. In *Proceedings of The 12th International Workshop on Semantic Evaluation*, pages 613–621, New Orleans, Louisiana. Association for Computational Linguistics.

Huong Dang, Kahyun Lee, Sam Henry, and Özlem Uzuner. 2020. Ensemble BERT for classifying medication-mentioning tweets. In *Proceedings of the Fifth Social Media Mining for Health Applications Workshop & Shared Task*, pages 37–41, Barcelona, Spain (Online). Association for Computational Linguistics.

Jacob Devlin, Ming-Wei Chang, Kenton Lee, and Kristina Toutanova. 2019. BERT: Pre-training of Deep Bidirectional Transformers for Language Understanding. In *Proceedings of the 2019 Conference of the North American Chapter of the Association for Computational Linguistics: Human Language Technologies, Volume 1 (Long and Short Papers)*, pages 4171–4186.

Koustava Goswami, Priya Rani, Bharathi Raja Chakravarthi, Theodorus Fransen, and John P. McCrae. 2020. ULD@NUIG at SemEval-2020 task 9: Generative morphemes with an attention model for sentiment analysis in code-mixed text. In *Proceedings of the Fourteenth Workshop on Semantic Evaluation*, pages 968–974, Barcelona (online). International Committee for Computational Linguistics.

Barry L Kalman and Stan C Kwasny. 1992. Why tanh: choosing a sigmoidal function. In *[Proceedings 1992] IJCNN International Joint Conference on Neural Networks*, volume 4, pages 578–581. IEEE.

Diederik P. Kingma and Jimmy Ba. 2015. Adam: A Method for Stochastic Optimization. In *3rd International Conference on Learning Representations, ICLR 2015, San Diego, CA, USA, May 7-9, 2015, Conference Track Proceedings*.

Arjun Magge, Ari Klein, Ivan Flores, Ilseyar Alimova, Mohammed Ali Al-garadi, Antonio Miranda-Escalada, Zulfat Miftahutdinov, Eulàlia Farré-Maduell, Salvador Lima López, Juan M Banda, Karen O'Connor, Abeed Sarker, Elena Tutubalina, Martin Krallinger, Davy Weissenbacher, and Graciela Gonzalez-Hernandez. 2021. Overview of the Sixth Social Media Mining for Health Applications (# SMM4H) Shared Tasks at NAACL 2021. In *Proceedings of the Sixth Social Media Mining for Health Applications Workshop & Shared Task*.

Vinod Nair and Geoffrey E Hinton. 2010. Rectified Linear Units Improve Restricted Boltzmann Machines. In *Proceedings of the 27th international conference on machine learning (ICML-10)*, pages 807–814.

Iulia Turc, Ming-Wei Chang, Kenton Lee, and Kristina Toutanova. 2019. Well-read students learn better: On the importance of pre-training compact models. *arXiv e-prints*, pages arXiv–1908.

Author Index

Association for Computational Linguistics
209 N. Eighth Street
Stroudsburg, Pennsylvania 18360

ISBN 978-1-7138-3029-0

CURRAN ASSOCIATES INC.
proceedings.com